ESTIMATING ABUNDANCE OF AFRICAN WILDLIFE
An Aid to Adaptive Management

ESTIMATING ABUNDANCE OF AFRICAN WILDLIFE

An Aid to Adaptive Management

by

Hugo Jachmann
Environmental Council of Zambia
Wildlife Resource Monitoring Unit
Lusaka, Zambia

KLUWER ACADEMIC PUBLISHERS
Boston / Dordrecht / London

Distributors for North, Central and South America:
Kluwer Academic Publishers
101 Philip Drive
Assinippi Park
Norwell, Massachusetts 02061 USA
Telephone (781) 871-6600
Fax (781) 681-9045
E-Mail <kluwer@wkap.com>

Distributors for all other countries:
Kluwer Academic Publishers Group
Distribution Centre
Post Office Box 322
3300 AH Dordrecht, THE NETHERLANDS
Telephone 31 78 6392 392
Fax 31 78 6546 474
E-Mail <services@wkap.nl>

Electronic Services <http://www.wkap.nl>

Library of Congress Cataloging-in-Publication Data

Jachmann, Hugo.
Estimating abundance of African Wildlife : an aid to adaptive management / by Hugo Jachmann
p. cm.
Includes bibliographical references (p.).
ISBN 0-7923-7959-4
1. Wildlife management—Africa. 2. Animal populations—Africa—Measurement. I. Title.

SK571 lJ33 2001
639.9'3'096—dc21

00-67107

Printed on acid-free paper. Printed in the United States of America

The Publisher offers discounts on this book for course use and bulk purchases. For further information, send email to <customer.service@wkap.com>.

TABLE OF CONTENTS

LIST OF FIGURES

LIST OF TABLES

LIST OF BOXES

FOREWORD

Information on animal abundance is vital to proper wildlife management, whether aimed at sustainable use, bio-diversity conservation or non-invasive tourism. Hugo Jachmann has worked as a wildlife ecologist in many African countries, and experimented with numerous techniques for estimating abundance of wildlife. In this manual he shares the lessons gained over more than 20 years of practical experience. Much of this time was spent in Zambia, where he first worked for the Luangwa Integrated Resource Development Project in the central Luangwa Valley. At the time of writing he holds the position of Chief Technical Advisor for the Wildlife Resource Monitoring Unit (WRMU) under the Environmental Council of Zambia (ECZ). In this post, he has identified a need for a comprehensive manual describing the various counting techniques, illustrated with simple examples from the field. Because of his 9 years experience in Zambia, many of the field cases originate from this country. This will make the manual especially useful for ECZ, and the recently created Zambia Wildlife Authority (ZAWA). Although the field work described here covers a period of more than 20 years, part of this manual was written whilst working for the WRMU.

The WRMU is a project of the Zambian Government, with financial and technical assistance from the Royal Netherlands Government, through its Embassy in Lusaka. The project aims at strengthening national capacity to manage the wildlife resource in an efficient and sustainable manner. Its three activity areas are to audit wildlife monitoring programs of other Zambian institutions, among others ZAWA, to provide training and monitoring assistance, and to produce a series of manuals. The first manual in this series describes the monitoring of illegal wildlife use and law enforcement (Jachmann, 1998), while the second manual gives a brief synopsis of Zambia's wildlife resources and its ecology (Jachmann, 2000).

Zambia provides a rich context for a work on wildlife monitoring. It has a broad variety of wildlife, which forms the prime resource base for tourism. Development of tourism, however, is seriously hampered because of the continuing threat to wildlife in many parts of the country. Although wildlife depletion was identified as one of the main environmental problems facing the country in the early 1990s, recent wildlife policies put emphasis on the commercial potential of wildlife utilisation to benefit the country and local communities. Rehabilitation of depleted areas and maintaining a balance between conservation and exploitation will be a major challenge for ECZ and ZAWA. Both are new institutions, without adequate capacity to meet this challenge. The development of field manuals such as this book and the sharing of experiences gained by persons like Hugo Jachmann will significantly contribute to building our national capacity. We hope colleagues working in the same field beyond our borders will also benefit from this manual, both utilising and building upon the practical techniques described.

James Simoko Phiri
Director
Environmental Council of Zambia

September, 2000

PREFACE

Over the past 23 years I have worked in many different locations on the African continent, applying numerous techniques for counting animals, ranging from down-to-earth dung to experiments in the skies. In between different posts there was always ample time to visit libraries in the Netherlands, to catch up with the latest developments. However, when starting with the Wildlife Resource Monitoring Unit in 1998, it became clear that local students, ecologists and wildlife managers were never in a position to visit well-stocked libraries. Furthermore, the methodology for counting animals is scattered amongst hundreds of scientific journals that are usually not available, while many of the more recent review papers focus on mathematics. This diffuseness of the literature, and its inaccessibility on the African continent, has frustrated many students and ecologists. Therefore, the objective in writing this book is an attempt to provide a text that can be used in undergraduate courses in wildlife management, and as a field manual for ecologists and wildlife managers. The book describes the various methods that are commonly used, illustrated with examples from the field. It should not be considered a complete work on counting animals, because the literature is too vast. I have only selected those techniques and analyses that I feel are immediately relevant. Because I am trained as an ecologist, with only basic mathematical knowledge, I have mainly chosen methods that are easily understood and applied in the field. Thus, the only expertise assumed in the reader is in basic statistics and some algebra.

The book is constructed in five interlocking parts. The first part provides a contracted overview of techniques, and an introduction to sample theory and basic statistics, illustrated with simple hypothetical examples (Chapters 1 & 2). This part may be of little interest to the advanced reader. The second part discusses direct techniques, divided into ground counts (Chapters 3 & 4) and aerial counts (Chapters 5 & 6). Starting with Chapter 4, concepts and techniques become more sophisticated, illustrated with examples from the field. The next part launches into a number of chapters providing an overview of indirect counts, starting with a brief description of mark/recapture, radio-tracking, DNA-analysis and index techniques (Chapters 7 & 8), and ending with a large sub-section on dropping counts and footprint measurements (Chapters 9 &10). Part IV describes briefly the analysis of distribution data and population trends (Chapter 11), and the last part discusses the sequence of decisions by which a technique is chosen for a particular set of conditions (Chapter 12).

In wildlife counts, the fundamental concept is total counts versus sample counts, where techniques are modified to suit either ground or aerial approaches. Because of the pronounced difference in platforms and the practical implications, I have chosen to discuss separately these two approaches.

Although the various techniques have been illustrated with examples drawn from a wide range of species and African habitats, I have to admit that there is a heavy concentration on elephants, not only because they are a fascinating and threatened species, but also because I have studied them for many years.

I would like to thank Richard Bell, Janice May and Rolf Lindholm for their constructive comments on some of the chapters. Furthermore, this manual has greatly benefited from numerous discussions with Richard Bell on the functions of monitoring wildlife populations and the various techniques for estimating wildlife abundance. I am immensely grateful to Richard Barnes, who likes to refer to himself as "malicious reviewer", but who reviewed and criticized every chapter and thereby contributed significantly to shaping this book. My sincere thanks to the staff of the Royal Netherlands Embassy in Lusaka, the Ministry of Environment and Natural Resources, the Director of the Environmental Council of Zambia, Mr. J. S. Phiri, and the staff of the Wildlife Resource Monitoring Unit for their continuous support throughout this project. Thanks to Debbie Peake from Safari Services Limited in Maun, Botswana, for providing me with the data on trophy measurements. Many thanks to Siân Taylor of Fergan Ltd. for typesetting the book and Fay Gadsden for compiling the index and proofing the final version. Last, but definitely not least, my gratitude goes to my wife and my children for their support, and above all their understanding, and simply for putting up with me.

Most of the field examples were collected over a period of more than 20 years. However, putting it all together into this manual was done whilst working for the Wildlife Resource Monitoring Unit at the Environmental Council of Zambia, a project financed by the Royal Netherlands Government, through its Embassy in Lusaka.

Hugo Jachmann

Environmental Council of Zambia
Wildlife Resource Monitoring Unit
Lusaka, Zambia

PART I

INTRODUCTION

Chapter 1

INTRODUCTION TO ESTIMATING WILDLIFE ABUNDANCE

Estimating animal abundance is central to sound management and is an important part of most wildlife research programs. Counting animals is a challenging task that will take you to many fascinating places that you would not otherwise see. I myself have spent many years in the bush trying to figure out how many animals were in my survey areas, and I hope that this manual will help you to achieve the level of satisfaction that I have gained from these studies.

The most obvious approach is simply to physically count all the animals in an area. This was attempted in the early days of aerial counting in Africa. However, with a few exceptions, this has now been abandoned as impractical and too costly. This change came in 1969, when George Jolly published his classic paper in which he set out the basic statistical principles for aerial sample counts (Jolly, 1969). The next landmark publication was the African Wildlife Foundation Handbook *Counting Animals*, which focussed on aerial techniques, but also contained a brief description of developments in ground-count methods (Norton-Griffiths, 1975). Since the 1970s, new techniques have been developed and old ones improved. Today's field worker has the choice of many different methods and analysis systems for the varied species, field conditions and objectives. However, the descriptions of these techniques are found in diverse scientific journals. Such journals are often inaccessible to African field workers and students, because the nearest internet connection and library may be light-years away. An additional problem with some of the newer techniques has been that complex mathematics and unfriendly computer software have served to frustrate or discourage the intended user.

In an attempt to fill the gap, this manual presents students, ecologists and wildlife managers with an introduction to the subject of animal counting. It goes on to describe the practical methodology of various techniques for assessing abundance of large African mammals. The statistical concept of sample theory is illustrated with simple examples before being developed further with a presentation of line-transect techniques. This section ends with a discussion of a simple short-cut alternative to using line-transect mathematics and software, called the mean perpendicular distance method. We continue with sections on aerial techniques, indirect methods, such as index and dropping counts, and the analysis of spatial distribution data and population trends. The final chapter discusses the selection of appropriate techniques for a particular set of conditions. Many of the techniques are illustrated with examples from the field, with the concentration of Zambian cases reflecting the time I have spent here. Although the manual describes techniques for counting African wildlife under conditions that prevail locally, with minor modifications, most techniques can be applied to large mammals found elsewhere.

Before getting too involved with the detail of the manual in an attempt to answer the intriguing question of how many animals there are in your area, you should first ask yourself why you need to count them. This may seem trivial, but the reasons for counting will give many clues as to the technique and the effort that should be employed, and therefore the funding required. Each reason requires different types of data, in terms of precision and accuracy. For example, regulation of consumptive use of wildlife needs accurate data, collected with procedures that minimise bias. Analysis of population trends requires precise estimates, which are obtained by rigid standardisation of methodology, and by taking large samples at high cost. However, when counts are designed to evaluate law-enforcement effort and efficiency for instance, it is often not necessary to formulate clear objectives. Here we can set the targets in terms of 'limits to acceptable change', and data collection may be done with little effort at low cost. Different field situations and study objects will also determine which techniques can be applied. Therefore, the technique and level of effort need to be specific to each situation and to each reason for counting.

With some of the techniques the general tendency is to produce underestimates. This, however, is unlikely to cause serious problems in wildlife management. Usually, the results of actions based on animal counts themselves provide means of improving the technique. This type of adjustment of both management actions and counting methodology is known as 'adaptive management'. Because of the complex nature of the systems studied and shortcomings in the techniques applied, adaptive management is a fundamental feature of successful wildlife management (Bell and McShane-Caluzi, 1985; Caughley and Sinclair, 1994).

1.1 A Brief Overview of Techniques

Wildlife counting techniques fall into two broad classes (Figure 1.1). The first comprises those techniques where the animals themselves are counted and where the entire population should be visible to the observers; these are ***Direct Counts***. The second class includes those techniques where none or only a proportion of the population is visible to the observers at any one time, or where indicators of the presence of the animals are used to determine their abundance; these are ***Indirect Counts.***

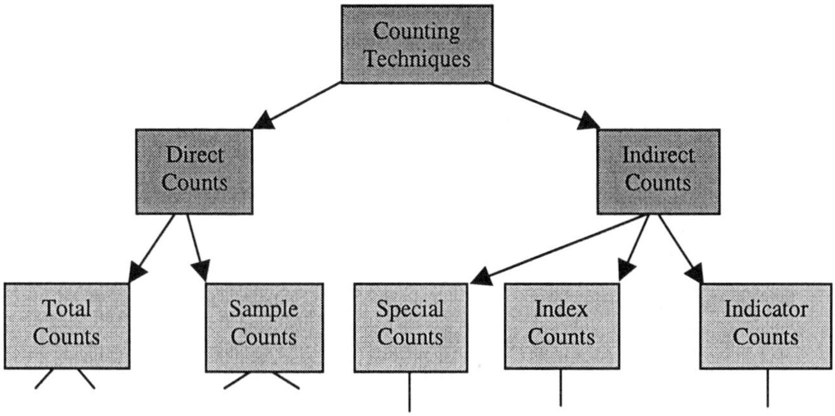

Figure 1.1: Techniques for estimating wildlife abundance.

Indirect counts can be subdivided into techniques that use the animals themselves to provide estimates of absolute densities (***Special Counts***), or indices of animal abundance (***Index Counts***), and techniques that use indicators of the presence of animals, such as droppings or footprints, to provide estimates of absolute densities, or indices of abundance (***Indicator Counts***), (Figure 1.1).

Direct counts can be subdivided into ***Total Counts or Censuses*** and ***Sample Counts or Surveys*** (Figure 1.1). In a total count, the entire census area is searched, and it is assumed that all animals have been seen and recorded. In a sample count, only part of the survey area is searched and counted. The number of animals counted in the sample area is then extrapolated to cover the entire survey area. Both total counts and sample counts can either be carried out from the ***Air*** or from the ***Ground***, which includes ***Water***. The basic principles for total counts from the air and those on the ground are similar, and the same applies to sample counts.

Both total and sample counts on the ground can be divided into ***Foot Counts*** and ***Vehicle Counts***. Sample counts by vehicle can be divided into ***Terrain Counts***, using the same principles as foot counts and used in wide open country, and ***Road Counts***, whereby the exercise is restricted to the existing road-system, usually applied to areas with high vegetation density and rugged terrain. Counts on the ground, whether carried out on foot or from a vehicle, can be ***Stratified*** or ***Non-Stratified***. Stratification means dividing the census or survey area into smaller areas called strata, each with a more or less homogenous density of the species to be counted. Sampling effort in the individual strata should be allocated on the basis of the density of the target species and the size of the stratum. The results of the independent samples from each of the strata can be combined, thereby improving the precision of the final estimate. Alternatively, the sampling effort can be the same for the entire area (non-stratified). The sample units can be placed at ***Random*** (random sampling) or ***Systematically*** (systematic sampling), (Figure 1.2).

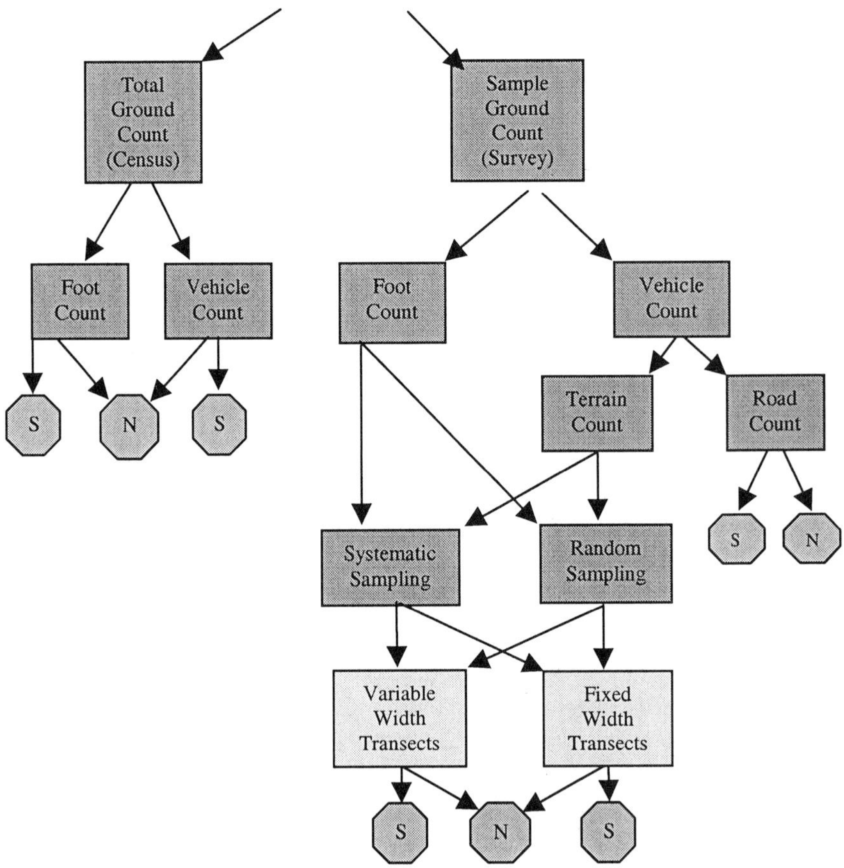

Figure 1.2: Techniques for counting wildlife on the ground (S=Stratified and N=Non-Stratified).

Ground counts may either use ***Fixed Width Transects*** or ***Variable Width Transects*** (strips with a fixed or variable width). A variety of techniques exists for variable-width transects, of which the line-transect technique is the most advanced and possibly the most reliable method. Generally, the methodology for vehicle counts is similar to that used for foot counts.

A special case of a total count on the ground uses individual recognition of all the animals in a particular population. This technique can be applied to a large number of species, such as elephant, giraffe, rhino, most of the predators and some of the more conspicuous antelopes. A major disadvantage is that the technique is time-consuming and therefore expensive.

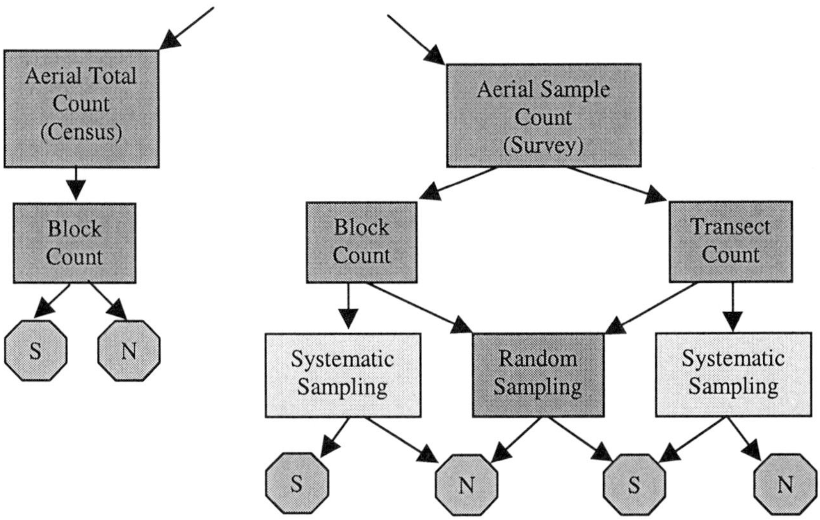

Figure 1.3: Techniques for counting wildlife from the air (S=Stratified and N=Non-Stratified).

Aerial sample counts use ***Blocks*** (rectangular areas) or transects as the sample units (Figure 1.3), either selected at random, or spaced at regular intervals. When blocks are used as the sample unit, the entire area of each individual block is searched and counted. Aerial total counts use blocks as the counting unit, which makes a total count similar to a sample block count, but in this case blocks cover the whole census zone.

Both in aerial total counts and in aerial sample counts, the census area and the survey area respectively can be stratified or non-stratified (Figure 1.3).

In certain special cases, line transects are used for aerial sample counts. However, this approach presents potential pitfalls. Without going into detail, it will suffice to state that several important assumptions are violated when using an aircraft for line-transect counts. Therefore, in this manual we will restrict discussion of line-transect methodology to its application on the ground.

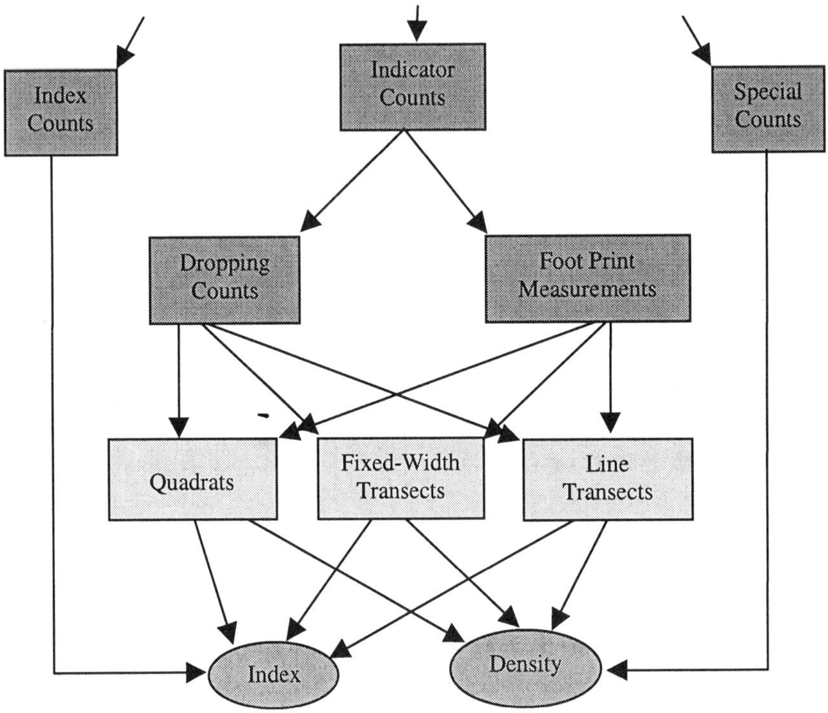

Figure 1.4: Indirect techniques for counting wildlife.

When no direct method is suited to a particular set of conditions, indirect methods must be used. This is the case when the animals are almost invisible to the observers, such as in the rain-forests of Central and West Africa. Also, when only a proportion of the animals are visible to the observers, which is frequently the case with large predators in most types of habitat, indirect methods should be used. Other arguments for using indirect methods are the relative cost-efficiency compared to direct methods, and sometimes the superior data quality, for instance elephant dropping counts in relatively small conservation areas (Jachmann, 1991a).

Numerous special and index techniques exist for counting wildlife, of which the following three methods are frequently used in practice:

-**Special Technique**; the *Capture/Recapture* method, where a known number of animals of a particular species is marked in some way and then released. Theoretically, the proportion of unmarked to marked animals in future catches can be used to estimate the total number of animals in the population under study. Unfortunately, the assumptions on which this method is based are difficult to satisfy in field situations and therefore it should be used with caution.

-**Index Techniques**; *Observations* made during *law-enforcement foot patrols*, corrected for effort (such as distance walked), can be used as indices of abundance (*Catch per Unit Effort*).

-*Hunting Success*, in terms of effort required to obtain a suitable individual of a particular species (e.g. time required to obtain a particular trophy size) can be used as an index of abundance (*Catch per Unit Effort*).

Most of these techniques will be dealt with in more detail at a later stage.

-***Indicator Counts*** can be subdivided into ***Dropping Counts*** and ***Foot Print Measurements*** (Figure 1.4). Dropping counts have been used to estimate the abundance of a large array of animal species ((Neff, 1968), but on the African continent the technique is mainly applied to estimating elephant abundance (Jachmann and Bell, 1979, 1984; Jachmann, 1988a & b, 1991a & b; Barnes and Jensen, 1987; Barnes, *et al.*, 1995; Barnes, *et al.*, 1997a). In certain special cases, in combination with foot print measurements, they are used for total-counting rhino in areas with low rhino densities (Jachmann, 1984a). Dropping counts can be used effectively for elephants and rhinos for three reasons. First the large size of the droppings make them readily visible. Second, the droppings are visible for long periods. Third, ample information is available in the literature on parameters required for analysis, such as seasonal defecation rates and seasonal faecal decay rates for elephant dung. The use of foot-print measurements to determine abundance is mainly limited to elephants, although sometimes used for rhinos, because the size and therefore the weight of the animals renders foot prints visible in many different types of terrain. However, the foot-print method can be applied only in areas with very low densities of elephants and/or rhinos, to enable cluster analysis of foot-print measurements of the individual animals.

The usual sample units for indicator counts are ***Quadrats*** (a square or rectangular area), fixed-width transects or line-transects (Figure 1.4).

1.2 Selection of a Technique

The different factors that influence the selection of a counting technique for a particular set of conditions will now be discussed briefly. Although this topic will be discussed in greater detail in Chapter 12 at the end of this manual, the reader needs to be introduced to the subject early on.

1. *The management objective*: The management objective depends upon the area in question and the species to be monitored. In conservation areas with non-consumptive utilisation of wildlife, such as national parks, the main objective is to maintain the integrity of the system, safeguarding it against loss of bio-diversity. Information on general trends of certain key species will be sufficient as an indicator of the state of the system and as feedback for law-enforcement and other types of management. General trends require periodic monitoring, to produce population estimates that are precise (*low variance; see section 2.2*), but not necessarily very accurate *(approaches the true number of animals within a population; see section 2.3*). In conservation areas with some form of consumptive utilisation (Game Management Areas (GMAs) in Zambia), such as safari hunting and culling operations, the objective will be to maintain economically important species at predetermined densities. Here, counts must be accurate, but preferably also precise, enabling time-series analyses and the calculation of annual off-take quotas for target species. Also, monitoring needs may vary by species. For example, in Zambia, elephants cannot be hunted, but occur in a number of GMAs and Open Areas. International legislation with regard to elephant conservation may dictate periodic counts that are unrelated to area-specific objectives.

2. *The topography of the area*: In mountainous and hilly country, aerial count techniques are difficult and results will be biased by problems of height control. Ground count techniques, direct or indirect, will be the appropriate alternative.

3. *The vegetation density*: Dense vegetation limits visibility for most techniques, resulting in underestimates. In many cases this problem can be partly avoided by using variable-width transects, such as line-transects. The disadvantage of line-transect methodology is that it is labour-intensive and therefore expensive.

Alternatively, indirect techniques, such as indicator methods (dropping counts and footprint measurements) can be used for elephants and rhinos, or patrol observations, road counts and radio telemetry for most other species.

4. *The distribution of the target population*: The precision of a sample estimate, obtained through aerial or ground techniques, is influenced by the density and distribution of the population being sampled. If the population is sparse and clumped, the sampling error (*see section 2.2*) is large. Index and indicator techniques may be appropriate alternatives.

5. *The species to be counted*: Some species can be counted by several methods, but others are difficult to count by any method, because of behaviour, body size, average group size and colouring. As an example, the greater kudu may seem conspicuous, but it is always under-counted by aerial techniques because it hides in thick vegetation during part of the day. Elephants on the other hand, can be counted with several techniques. Although aerial techniques are used for many species throughout the continent, it is probably safe to conclude that they only produce reliable results for zebra, buffalo, lechwe, wildebeest and elephant when large herds are photographed. For species that are hunted, indirect techniques, such as trophy quality, produce information of an adequate quality at a low cost for adjusting annual off-take quotas. Special aerial surveys, using slow and low-flying micro-light aircraft or helicopters, in combination with photography, apply to crocodile, hippo and sitatunga. The majority of ungulate species can be counted on line-transects, but again, in hunting areas, trophy quality provides adequate information in most cases. Puku, reedbuck and oribi, however, always follow a clumped distribution, concentrating in dambos and on floodplains. The only appropriate method to count these antelopes is a total ground count, while trophy quality data may be appropriate in hunting areas. Predators, such as lion, leopard, cheetah and wild dog require indirect methods, such as trophy quality, patrol observations or special studies, such as baiting of lions, combined with individual recognition. A detailed summary is provided in Chapter 12 (*see Table 12.1*).

6. *The budget*: At first sight this factor may seem trivial and not directly related to any of the factors described above. However, in

practice it is an important factor determining the technique for a particular situation. Direct wildlife counts are expensive and consequently not an affordable option for many stakeholders. Cheaper methods may have to be used. Certain indirect methods are cost-efficient, and if used consistently will produce adequate information to generate the correct management decisions.

7. *Logistical factors*: The availability of adequately trained staff, transport, equipment and analytical facilities are important factors, independent of the technique applied.

Chapter 2

INTERPRETATION OF POPULATION ESTIMATES

In the late 1950s, scientists and wildlife managers began using light aircraft to count animals on large tracts of African savanna rangeland. At first they were doing total counts, but then they realised that it would be far more cost-efficient to use some form of sampling. It was also becoming apparent that for management purposes *trends* in wildlife populations were possibly more important than *total numbers*. To the layman, however, it is important that we should try to count all the animals present in an area, and this is where the problem arises. A sample count provides an estimate of abundance, rather than total numbers.

Although total numbers are usually readily accepted, estimates of abundance are open to abuse. The most common form of abuse is misinterpretation of the statistics describing the precision of the population estimate. Sometimes misinterpretation is caused by the critic's limited knowledge of statistics, but in some cases critics deliberately misinterpret or manipulate wildlife counts so as to mislead their audience. To minimise misunderstanding of the statistics describing a population estimate, we will start with a brief introduction to the basic principles of statistics and the interpretation of sample counts.

2.1 Population Estimate

Direct wildlife counts are either *total* counts or *sample* counts. In sample counts only part of the survey area is covered, either using transects (strips with a fixed width or strips with a variable width) or blocks (areas with a rectangular shape) as the sample units. The results are then extrapolated to cover the entire survey area. Using the data collected during a sample count, the size of the population can be *estimated.* This estimate may be close to the *true number of animals* in the survey area, or it may be far from the true number. In most cases we do not know the true number of animals. When the target population is not evenly distributed over the survey area, the results of the count will partly depend on the number and placement of the sample units selected.

To illustrate this, we use a simplified hypothetical example, where a small survey area contains 10 elephants. These elephants spend most of their time in the western part of the survey area. An aerial sample count, with 5 parallel transects placed at regular intervals (*systematic placement*), covers the entire survey area with a sampling intensity of 5%. During the survey, only 1 elephant is counted in one transect (Figure 2.1). Extrapolation of the sample to cover the entire survey area gives an estimate of (100/5) x 1 = 20 elephants. This estimate is twice the true number and will have a large sampling error, because only 1 out of a small sample of 5 transects contained one elephant. The results of the first survey showed management that the eastern half of the survey area contains no elephants. For the second aerial sample count, 5 transects are placed at regular intervals, covering the western half of the survey area only, with a sampling intensity of 10%. Again, only 1 transect contains 1 elephant (Figure 2.1). Extrapolation gives an estimate of (100/10) x 1 = 10 elephants. Although this estimate gives the true number present in the area, it has a large sampling error. To obtain an estimate that approaches the true number and has a small sampling error, the number of transects covering the western half of the survey area should be increased to give a higher sampling intensity. The subject of sampling error will be discussed in the next section.

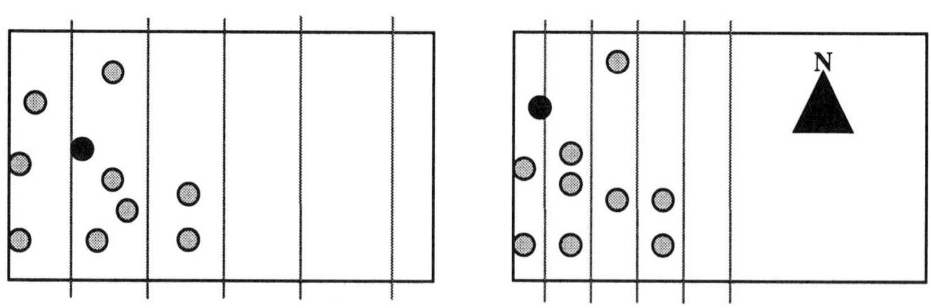

Figure 2.1: Diagram of hypothetical example of a small survey area with 10 elephants. Five transects placed at regular intervals to cover the entire survey area with a 5% sampling intensity (left), and 5 transects placed at regular intervals in the western half of the survey area with a 10% sampling intensity (right), both return one elephant sighting (black dot). The estimate for the 10% sample gives the true number, whereas the estimate for the 5% sample is twice the true number.

2.2 Variance, Standard Error and Confidence Limits

In sample counts, when using transects or blocks as the sample units, we can calculate a population estimate for each individual unit. The estimates from individual sample units will cluster about the mean value, which is the sum of *n* estimates divided by *n*. If the estimates are tightly clustered around the mean, it means there is little variation between sample units (Figure 2.2). This variation is measured by the *Variance.* Thus, the variation in the population gives the sampling error, which in turn is measured by the variance. A small variance in a sample count implies that the target population is more or less evenly distributed over the survey area. If the estimates are widely spread around the mean, the variance is high (Figure 2.2). A high variance means that the target population is unevenly distributed over the survey area.

The variance is the sum of the square of the error from the mean of each sample unit, divided by the number of units or the sample size (*n*). However, for small samples (<30), dividing by *n* tends to underestimate the population variance. Here, a better estimate is provided by dividing by $n - 1$. For large samples (n>30), however, dividing by *n* or by $n - 1$ produces similar results. The deviation from the mean of the individual sample units (*Standard Deviation*) is squared to avoid negative values. The standard deviation of a set of independent estimates of the same population is a measure of *precision* of any of the independent estimates. It can be estimated from the quadrat counts of a single survey. Then it is known as the *Standard Error.* Thus, with animal counts standard error is used instead of standard deviation. The standard error is obtained by taking the square root of the variance (*see Box 2.1*).

The standard error provides additional information about a frequency distribution of a data set. Suppose we have 100 observations (*n*), with a mean of 5.96 (x_m) and a standard error (*SE*) of 2.03. A smooth curve is drawn, connecting the midpoints of the tops of the rectangles of the histogram (Figure 2.3). If the resulting curve is approximately bell-shaped (the data are approximately normally distributed), then about 68% of the data will lie in the interval from x_m – SE to x_m + SE, about 95% of the data (*Confidence Limits*) will lie in the interval from x_m – 2SE to x_m + 2SE, and most of the data will lie in the interval from x_m – 3SE to x_m + 3SE (Figure 2.3).

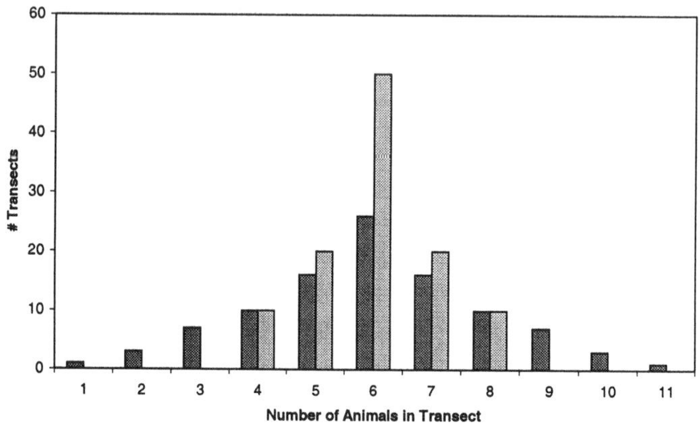

Figure 2.2: A simple illustration of variance. In the first sample (light grey) the variance is low with the estimates clustered around the mean. In the second sample (dark grey) the variance is high, with the estimates widely spread around the mean. The mean is the same in each case.

The *Confidence Interval* or *Confidence Limits* are usually expressed as a range above and below the population estimate, i.e. the *Lower Limit* and the *Upper Limit.* There are two alternative interpretations of the concept "confidence limits". Using the first interpretation, the 95% confidence limits simply mean that with a 95% certainty the true population size lies within this interval. The alternative interpretation is that with repeated sampling, 95% of the resulting population estimates will fall within this range (*repeatability*).

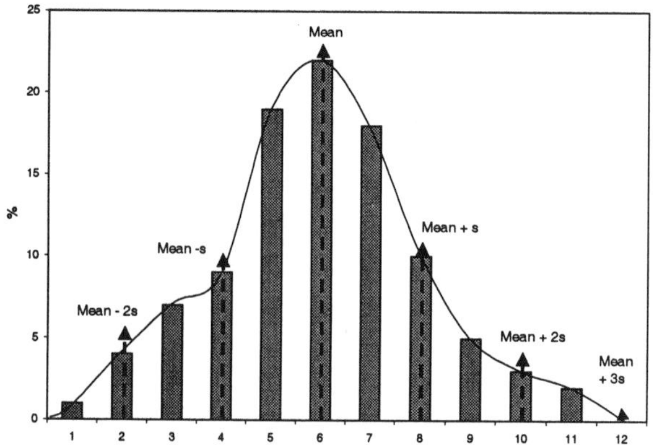

Figure 2.3: Frequency distribution of 100 observations. About 68% of the data lie within the interval mean – SE to mean + SE, about 95% within the interval mean –2SE to mean + 2SE, and most data within the interval mean –3SE to mean + 3SE.

The 95% confidence limits can be estimated by multiplying the standard error by a statistic t. The value of this factor depends on the number of observations, but when *n* is large (n>30), t is 1.960 (i.e. the interval between $x_m - 2SE$ to $x_m + 2SE$; *see Figure 2.3*).

In biological studies, 95% confidence limits are the most common measure of precision. However, in particular situations it may be required to provide a lower or higher level of precision and then you may wish to calculate the 90% or 99% confidence limits respectively. When *n* is large (n>30) t is 1.645 for 90% limits and t is 2.576 for 99% limits.

A population estimate with a small variance and therefore a small standard error and narrow confidence limits is a precise estimate.

To illustrate this, we use a highly simplified hypothetical example with a survey area of 100 km x 100 km (10,000 km²) containing 1,000 elephants. During the dry season, the majority of elephants are concentrated in an area of approximately 2,000 km², in the proximity of the only remaining water source. During the wet season, the elephants are evenly spread out over the survey area with a mean group size of between 4 and 5. The resident biologist, who has no clue about the size and the distribution of the population, decides to use aerial sample counting in both the dry and wet seasons to estimate elephant numbers. The budget that is available for this exercise allows placement of 20 parallel transects, each 100 km long. Transect width is 200 m on either side of the aircraft. The size of the *Sample Area* is the number of transects x width x length = 20 x 0.4 km x 100 km = 800 km². The percentage coverage of the study area, or the sampling intensity, is (800 km²/10,000 km²) x 100 = 8%. The total number of possible transects in the population (*Population of Sample Units*) is the length of the survey area (*Base-line*) / transect width = 100 km/0.4 km = 250.

During the dry-season count he observes a total of 100 elephants in only 2 out of 20 transects (2 transects with 50 elephants each and 18 transects without elephants). The wet-season count gives a total of 80 elephants in 18 out of 20 transects (10 transects with 4 elephants, 8 transects with 5 elephants and 2 transects without elephants). Using the formulae provided in Box 2.1, we can calculate the mean number of elephants and the standard error for the transects that were sampled, for both seasons:

The dry-season mean is:

$$x_m = \frac{50 + 50 + (18 \times 0)}{20} = 5$$

The standard error is:

$$SE = \sqrt{\frac{(5 - 50)^2 + (5 - 50)^2 + (5 - 0)^2 + \ldots\ldots}{20 - 1}} = 15.39$$

%SE = (15.39/5) x 100 = 307.80% (i.e. the SE expressed as a percentage of the mean)

The dry-season population estimate is:

250 x 5 = **1,250 elephants**

The wet-season mean is:

$$x_m = \frac{(10 \times 4) + (8 \times 5) + (2 \times 0)}{20} = 4$$

The standard error is:

$$SE = \sqrt{\frac{(4 - 4)^2 + \ldots + (4 - 5)^2 \ldots + (4 - 0)^2 + (4 - 0)^2}{20 - 1}} = 1.45$$

%SE = (1.45/4) x 100 = 36.25%

The wet-season population estimate is:

250 x 4 = **1,000 elephants**

The standard errors for the transects show that the dry-season estimate is very imprecise, whereas the wet-season estimate is relatively precise. In summary, due to an even distribution of the elephants over the study area, the wet-season count gave a low random variation between transects, resulting in a low sampling error and a relatively precise estimate. The clumped distribution of elephants during the dry season, however, when only a single water point was remaining, resulted in a large sampling error, and a very imprecise estimate.

Box 2.1: Formulae for mean, variance, standard error and confidence limits.

The mean or average (x_m) of *n* observations is given by:

$$x_m = \frac{\sum x}{n}$$ (The symbol $\sum$ means summation of all observations)

The variance of *n* observations with mean x_m is estimated by:

$$s^2 = \frac{\sum (x_m - x)^2}{n - 1}$$

The standard error of n observations with mean x_m is given by:

$$SE = \sqrt{s^2}$$

The 95% confidence limits are given by:

$t \times SE$ (If $n > 30$ then $t = 1.96$)

The three examples presented to illustrate the concepts of variance and standard error, one with low variance and two with high variance (Figures 2.2 & 2.3), are purely theoretical. The data follows a near normal distribution. Certain continuous biological data, such as stature of people, follows a normal distribution, which is a smooth symmetrical bell-shaped curve. The normal curve is one of the easiest to handle theoretically, and therefore, statistical analysis can be done with a minimum amount of computation. However, in practice, the distribution of count data has a different shape.

A field example, showing some of the results of an aerial survey carried out in the South Luangwa National Park in 1998, will demonstrate the typical shape of the distribution of count data (Figure 2.4). The South Luangwa National Park covers 9,050 km² in the east of Zambia, and is surveyed annually, using the same 43 parallel transects spaced at 2 minute (approximately 3.7 km) intervals. Examples are provided for three common species, elephants, which occur throughout most of the park, zebra, also with a more or less even distribution, and hartebeest, mostly occurring in the munga and miombo woodlands in the western parts of the park.

The frequency distribution of the numbers of elephants counted in 43 transects shows a smooth regular pattern, that is heavily skewed towards low values (Figure 2.4). This means that there were many transects in which only few or no elephants were seen. Nevertheless, elephants were seen over most of the park, with observations in 28 out of 42 transects. Zebras were seen in 23 transects, while hartebeest were seen in only 13 transects (Figure 2.4). Note how large numbers of hartebeest were seen in some transects, while most transects recorded no hartebeest. These distributions resulted in a precise population estimate for elephants, with a standard error of only 13.4%, a standard error for zebra of 20.9% and a standard error for hartebeest of 29.6% (Figure 2.4).

This example from the field shows that in practice, count data is usually heavily skewed towards zero values, that is, there are many transects in which no animals are seen. This usually leads to count data following a Poisson distribution, especially with small sample sizes (Figure 2.4). The Poisson distribution involves only one parameter (m), which can be estimated by calculating the mean of a sample with n measurements. The standard error can be approximated by taking the square root of m. The significance of this is purely statistical. For example, when comparing two data sets, both with a normal distribution, powerful parametric statistics, such as the t and F-tests, can be applied to test for significant differences between the means. Data following a Poisson distribution, however, require less robust non-parametric statistics, or a transformation to make the data follow a more or less normal distribution.

At this stage in the manual, the above example and the brief discussion of the distribution of count data serves three purposes. First, the example clearly demonstrates the close relationship between spatial distribution of animals and the precision of survey results. Second, it serves to show the reader that initially, simplified theoretical examples are required to explain some of the statistical concepts, but that data collected in the field are never this perfect, follow different distributions, and are therefore often more complicated to analyse. Third, to provide the reader with a brief introduction to one of the common distributions of count data, because this is an important integral part of the theory of some of the methods presented further on in the manual.

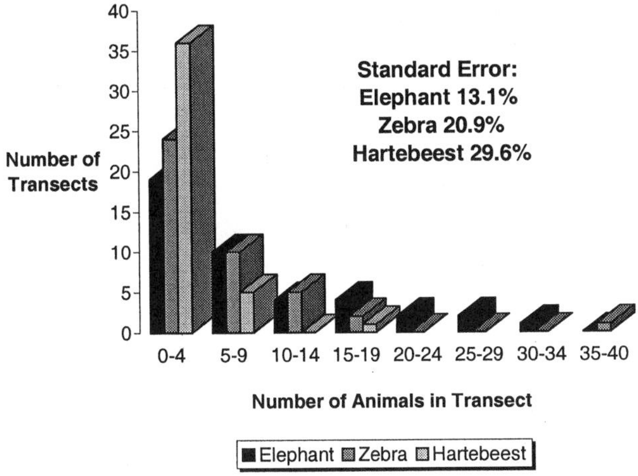

Figure 2.4: Three examples of the results of an aerial survey carried out in the South Luangwa National Park in 1998. The Poisson distributions for elephants, zebra and hartebeest observed in 43 parallel transects.

2.3 Bias and Accuracy

In addition to random variation, population estimates are also affected by errors that are more or less constant for a particular survey, called *biases*. For aerial techniques there are three common sources of biases: i) the limited visibility of the target population; ii) the observers' experience and their ability to concentrate; iii) the speed and height of the aircraft, and the width of the transects. In most cases these errors bias the final estimate downwards (*negative bias*), leading to underestimates. The section dealing with aerial counting techniques (*see Part IIb*) gives a detailed discussion of bias.

The random errors or variation in a sample determine the *precision* of the final population estimate, whereas the constant errors or biases influence the *accuracy* of a population estimate. An *accurate* estimate is one that approaches the true number of animals within the population under study. However, an accurate estimate may not be precise and may have wide confidence limits. Let us examine our previous example, where the wet-season elephant count gave a population estimate of 1,000 animals and the dry-season count an estimate of 1,250 animals, in an area with a true population of 1,000 elephants. Here, we can conclude that the wet-season estimate is accurate, while the low variation between sample units shows that it is also relatively precise. In contrast, the dry-season estimate is neither very accurate nor precise. The management objective for the area under consideration, and the species in question determine whether we require a precise or an accurate estimate, or both. Although we are never sure how accurate an estimate is, we can limit bias through a proper design.

In total counts, repeated counts of the same area will give different results, mainly due to changes in the biases inherent to the method used. In a total count, the main sources of bias are the proportion of the study area that is not searched and the proportion of the target population missed by the observers.

2.4 Stratification

Stratification of the study area is used to improve the precision of the final population estimate. Stratification involves dividing the study area into a number of separate areas, known as *strata*. Stratification may be done according to permanent topographical features, such as mountain ranges, rivers, etc., or in relation to more temporary aspects, such as seasonal concentrations of animals. Each stratum has a different, but more or less homogeneous density of the target population. An independent sample is taken from each stratum. The strata are sampled at different intensities.

Stratification will reduce the variance only if there are large differences between strata. Thus, information on the densities of the target species in the various strata should be obtained prior to a count, either through a reconnaissance survey, or by using data from previous counts.

When counting a species that follows an irregular distribution pattern with significant differences in densities between areas, stratification becomes a useful tool to improve the precision of the estimate. If, however, several species are counted during the same survey, and the distribution of these species is not strongly related, one effective stratification for all species will be difficult to find. In this case, stratification is not advised.

After the study area has been stratified, the sample size for each stratum should be determined. Allocation of sampling effort should be in relation to density and stratum size. It would be a waste of resources to expend much effort on large areas that contain few animals, because the results will have little influence on the accuracy and precision of the final estimate. Thus, the higher the density of the target species and the larger the area of the stratum, the higher the sampling effort. For a detailed mathematical explanation on gain in precision with stratification, see

Cochran (1977). The optimum allocation of sampling effort can be approximated by (C. Craig, unpublished; quoted in Martin *et al.*, 1992):

$$S_i = (P \times A \times a_i \times \sqrt{d_i}) / (\Sigma\, a_i\sqrt{d_i})$$

where:

S_i = area to be sampled in stratum i (km²),
P = sampling intensity (%),
A = total area of all strata combined (km²),
a_i = area of ith stratum (km²),
d_i = density of target species in ith stratum (animals/km²).

Our example of the wet and dry-season elephant counts conducted in the survey area of 10,000 km² illustrates the use of this formula. Although the wet-season count gave excellent results, the dry-season count was very imprecise due to the clumped distribution of the elephant population. Let us assume that the resident biologist has decided to stop aerial survey work during the wet season, mainly on account of the dangers involved when flying in unstable weather conditions. Although he does not know how accurate his dry-season estimate is, the precision was considered not acceptable. He decides to use the information of the previous surveys to stratify the area for the next dry-season count. He decides to place 40 parallel transects instead of 20. The distribution from the first dry-season count suggests two strata, one stratum of 2,000 km² with a density of 0.48 elephants per km² (stratum 1), and a second stratum of 8,000 km² with a density of 0.005 elephants per km² (stratum 2). The allocation of sampling effort can be approximated as follows:

Total survey area (A) is 10,000 km²,

Sampling intensity or area coverage (P) is {(40 x 100 x 0.4)/10,000} x 100 = 16%,

Total sample area (P x A) is 1,600 km²,

Area of stratum 1 (a_1) is 2,000 km²,

Area of stratum 2 (a_2) is 8,000 km²,

Elephant density in stratum 1 is 0.48/km²,

Elephant density in stratum 2 is 0.005/km²,

Stratum 1: S_1 = (1,600 x 2,000 x 0.69) / (560 + 1,380) = 1,138 km^2

Stratum 2: S_2 = (1,600 x 8,000 x 0.07) / (560 + 1,380) = 462 km^2

Placement of 40 transects; each 0.4 km x 100 km = 40 km^2, is as follows:

In stratum 1 (1,138/40) 28 transects will be placed, with the remaining (462/40) 12 in stratum 2.

Because most of the population is concentrated in a relatively small area, a suitable alternative would be to conduct a total count in stratum 1, and a low-intensity sample count in stratum 2.

The major disadvantage of a stratified count is that it collects less information on the spatial and temporal distribution of the target species. However, because this type of information is initially required to stratify the area, a stratified count is usually part of an ongoing monitoring program. Then, we can reword this by saying that a stratified count leads to less information on changes in spatial and temporal distribution. Changes in distribution patterns of key species may provide management with important feedback for law-enforcement operations, or other management-related activities. Also, the main advantage of a non-stratified approach is the option to examine factors that influence the spatial and temporal distribution of a single species or several species that were all counted during the same survey. For instance, the data on animal abundance can be related to rainfall patterns, to a soil or vegetation map or to disturbance factors such as illegal hunting (*see Chapter 11*).

2.5 Random Sampling versus Systematic Sampling

In the late 1960s, two aerial sampling techniques were developed, the stratified random sampling method, and the systematic reconnaissance flight technique. Both techniques used similar data-gathering methods and analysis procedures. The difference between the two techniques was in the type of information generated. The random sampling method gives a population estimate, but provides little information on animal distribution or habitat features. The systematic reconnaissance flight technique, or systematic aerial sample count, however, gives data on both numbers and distribution. Both techniques, only with minor adaptations, are still in use. Random and systematic sampling differ in the placement of transects. In both cases we start by dividing the study area into sample units, whereby the numbered transects run perpendicular to a base-line. In systematic sampling we simply keep the interval between consecutive transects the same, for instance every third transect. In random sampling, we can use a random number table to select transects. For random sampling, each sample unit has an equal chance of being chosen, each unit occurs once in a sample, and all possible combinations of sample units are equally likely to occur (*sampling without replacement*).

We have seen that less information is collected during a stratified survey. The same applies to random sampling as compared to systematic sampling. To illustrate this, we return to our example. If our biologist decided on a random sample count, the results would have provided some information on the spatial distribution of the elephant population, but not sufficient to properly stratify the study area for the following year. In our example, stratification was a necessity.

Generally, if the distribution of the target species is relatively homogeneous over the study area, there is little practical difference between random versus systematic sampling. Then, the variation is of a purely random nature. In practise, however, very few animals are evenly distributed. There is always some environmental factor influencing animal distribution, usually in the form of a trend across space. This indicates an advantage in systematic sampling. Thus, if there exists a gradient in the density of the target species, in addition to random variation, systematic sampling will produce more precise estimates. The only disadvantage of systematic sampling is that the *estimated* standard error is usually higher than the *true* standard error, but always lower than one obtained through random sampling. Therefore, the precision of any population estimate derived through systematic sampling should be considered slightly inflated.

2.6 Sample Count Theory

Sample count theory will now be developed further, using two hypothetical field examples, based on conditions commonly found in the Lupande GMA, Zambia. The survey area is located in the alluvial part of the Luangwa Valley and is 20 km long and 5 km wide (100 km^2). The first example describes a sample count from a true population of 200 common duiker (density is 2 duikers per km^2). The common duiker is predominantly a solitary and sedentary species, and is distributed evenly over the survey area, both on a temporal and on a spatial basis. In the Lupande GMA, the mean group size for common duiker is usually around 1.1 (range 1 to 2). The second example describes a sample count from a true population of 200 elephants (density is 2 elephants per km²). Elephants are social animals that occur in family units, kinship groups or entire clans, depending upon the season (that is on water and food availability) and disturbance factors such as poaching. In addition, elephants are usually on the move and consequently have a far more erratic spatial and temporal distribution pattern than the common duiker. In the Lupande GMA, in the dry season, the mean group size for elephants is 4.8 (range 1 to 16). The reason for counting both species is to obtain population estimates that are both accurate and precise, for purposes of setting off-take quotas for hunting.

The objective of a sample count is to estimate the number of a particular species by searching and counting a proportion of the survey area. That proportion is called the sample area. It must be representative of the whole survey area. Therefore, the survey area is divided into a number of discrete units known as sample units. These sample units may have the form of equally sized blocks or equally sized strips. They can be chosen randomly or systematically. The use of blocks has many disadvantages, which will be discussed in following chapters (*see 3.2 and 6.1.1*), while the disadvantages of random sampling were discussed above. Therefore, at this stage, the discussion will be limited to systematic strip sampling.

We have already mentioned that the common duiker is a solitary and sedentary species with an even distribution over the survey area. If this is true, each sample unit will contain approximately the same number of duikers and we will only need to count a small number of sample units to calculate our sample estimate for the whole area. Elephants on the other hand have a far more erratic distribution pattern. To obtain a sample estimate with the same level of precision, the number of sample units required will be much higher. However, for both species the larger the number of sample units, the more precise the estimate. In other words, there are many alternative estimates that can be obtained depending upon the size of the sample. The range of possible estimates is however much smaller for the common duiker than for elephants. The error of a sample estimate from the true number of animals is known as *sampling error*. Although any single sample estimate may be higher or lower than the true number of animals in that particular population, if you take a large number of samples of the same area and the same species, sampling errors will balance out and the mean of the sample estimates will equal the true population total.

Thus, the size of the sampling error in a particular sample depends on the variation in the number of animals of the target species counted in each of the units selected. This is measured by the sample variance. If the sample variance is small, numbers of animals counted in the different sample units were about the same, and the range of alternative estimates will be small. Obviously, if the sample variance is large, numbers of animals counted in the different sample units varied widely, and the range of alternative estimates will also be large. This range of alternative estimates is provided by the 95% confidence limits, which is a measure of its repeatability or the precision of the estimate. The formulae for calculating the mean, variance, standard error and 95% confidence limits are provided in Box 2.2 (Jolly, 1969). In the sample below, we will choose the number of strips to be sampled for both species.

On average, the visibility in the survey area is approximately 50 m on either side, giving a total strip width of 100 m. To limit the variation in the results, a large number of short strips is preferred to a small number of long strips. In our case this means that placement of strips will not be along the length of the survey area (20 km), but along its width (5 km). The total number of possible sample units in the survey area is 20 km divided by 0.1 km is 200 strips.

Box 2.2: Formulae for sample mean, sample estimate, sample variance, population variance, population standard error and 95% confidence limits (Jolly, 1969).

The sample mean ($\hat{y}$) of *n* observations is given by:

$$\hat{y} = \frac{\Sigma y_i}{n}$$

The sample estimate *Y* of *N* sample units is given by:

$$Y = \hat{y} \times N$$

The sample variance of *n* observations is estimated by:

$$s_y^2 = \frac{1}{n-1} \left\{ \Sigma y^2 - \frac{(\Sigma y)^2}{n} \right\}$$

The population variance of n observations in N sample units is estimated by:

$$Var(Y) = \frac{N(N-n)}{n} \times s_{y2}$$

The population standard error is given by:

$$SE(Y) = \sqrt{Var(Y)}$$

The 95% Confidence Limits are given by:

$$\pm t \times SE(Y)$$

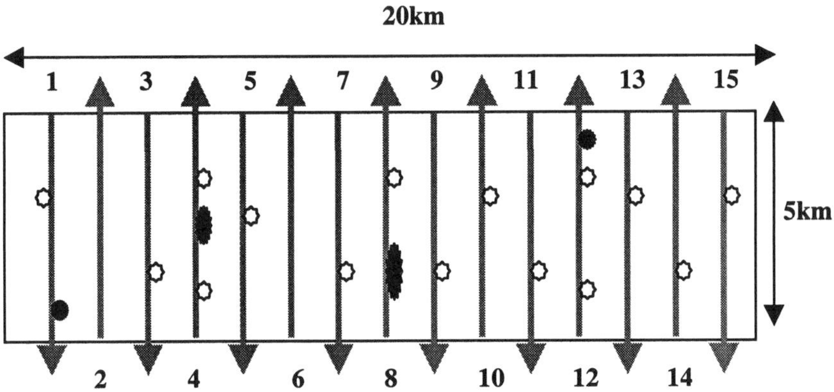

Figure 2.5: Diagram of hypothetical survey area of 20 km x 5 km, showing 15 sample units (indicating line of travel) out of a total population of 200 sample units. The sample area is the percentage of the survey area covered by the sample units or strips. Duikers counted are indicated with white stars, and elephants with black stars.

Common Duiker: We arbitrarily choose a sample size of 15 strips out of the total of 200 strips. Sampling intensity is (15/200) x 100 = 7.5%. The strips are placed systematically, with a 5-km long strip every 1.33 km (Figure 2.5). The results are as follows:

Strip Number:	1	2	3	4	5	6	7	8	9	10	11	12	13	14	15
Number of Duikers:	1	0	1	2	1	0	1	1	1	1	1	2	1	1	1

1-*Calculation of the Sample Estimate*:

Total number of sample strips; $N = 200$
Number of strips in the sample; $n = 15$

Number of duikers counted in the sample;
$\Sigma y = 1+1+2+1+1+1+1+1+1+2+1+1+1 = 15$

Sample mean; $$\hat{y} = \frac{\Sigma y}{n} = \frac{15}{15} = 1$$

Sample estimate: $Y = \hat{y} \times N = 1 \times 200 =$ **200 duikers**

2-*Calculation of the Sample Variance*:

$\Sigma y = 15$
$(\Sigma y)^2 = 225$
$\Sigma y^2 = 19$

Sample variance; $$s_{y^2} = \frac{1}{n-1}\left\{\Sigma y^2 - \frac{(\Sigma y)^2}{n}\right\} = \frac{1}{15-1}\left\{19 - \frac{225}{15}\right\} = 0.29$$

3-*Calculation of the Population Variance*:

Intensity of sampling (weighting factor); $$\frac{N(N-n)}{n} = 2{,}466.67$$

Population variance; $$Var(Y) = \frac{N(N-n)}{n} \times s_{y^2} = 2{,}467 \times 0.29 = 715.43$$

4-*Calculation of the Population Standard Error*:

Population standard error;
$SE(Y) = \sqrt{Var(Y)} = 26.75 = 26.75/200 \times 100 = 13.38\%$

5-*Calculation of the 95% Confidence Limits*:

95% Confidence limits;

CL(Y) = 2.145 x SE(Y) = 2.145 x 26.75 = 57.38 = 28.69%

In this case the sample estimate is equal to the true number of duiker in the population. Usually we will not know whether a sample estimate is accurate or not, but its precision will provide some information on its repeatability. In our case, the precision of the estimate can be expressed as: Y = 200 ± 13.38% duiker (95% CL 147 – 253), which is a very accurate and a very precise estimate obtained through a survey with an area coverage of only 7.5%. The accuracy and precision of the estimate are due to the solitary nature of the species and the high density and the even distribution of the population in this survey area. In terms of the objective of the count, the precision of the estimate is more than adequate to set an off-take quota for hunting purposes. To limit confounding differences in bias levels we should use the same design and procedures for surveys carried out in following years.

Elephant: If we repeat this exercise for a social species such as elephant, the outcome may be less accurate and certainly less precise (Figure 2.5):

Strip Number:	1	2	3	4	5	6	7	8	9	10	11	12	13	14	15
Number of Elephants:	1	0	0	6	0	0	0	11	0	0	0	2	0	0	0

Total number of sample strips; N = 200
Total number of strips in the sample; n = 15
Number of elephants counted in the sample; Σy = 20

Sample mean; $\hat{y} = \frac{20}{15} = 1.33$

Sample estimate; Y = 1.33 x 200 = 266.67 = **267 elephants**

Sample variance;

$\Sigma y = 20$

$(\Sigma y)^2 = 400$

$\Sigma y^2 = 162$

$$s_{y^2} = \frac{1}{14} \left\{ 162 - \frac{400}{15} \right\} = 9.67$$

Population variance; Var(Y) = 2,467 x 9.67 = 23,855.89

Population standard error; SE(Y) = √Var(Y) = 154.45 = 57.85%

95% Confidence limits; CL = 331.30 = ± 124.08%

The precision of the estimate can be expressed as:

Y = 267 ± 57.85% (95% CL 0 – 598)

We must conclude that in the case of elephants our sample was too small, and the estimate too imprecise for quota-setting purposes. Increasing our sample size will provide a more precise, and possibly a more accurate estimate. The alternative is to use another technique for estimating elephant abundance. However, because there are no previous hunting data, cost-efficient index techniques cannot be used (*see Chapter 8*), while ground count techniques that will produce an estimate of an adequate precision, such as line transects or dropping counts, may be too costly to deploy.

Let us examine the results of an aerial sample count with three times the area coverage, e.g. 3 x 7.5% = 22.5%, assuming the same spatial distribution of elephants. Forty-five strips are placed at regular intervals. The results are exactly the same as those of the first count, with three times the number of elephants (60), distributed with the same frequency (e.g. three times the results of the 15 strips covered in the first count). Thus, the sample mean and sample estimate are the same as in the first count ($\hat{y}$ = 1.33 and Y = 267 elephants), but the population variance, and therefore the precision of the estimate will improve.

Calculations are as follows:

$\Sigma y = 60$

$(\Sigma y)^2 = 3,600$

$\Sigma y^2 = 486$

Sample Variance;

$$s_y^2 = \frac{1}{44}\left\{486 - \frac{3,600}{45}\right\} = 9.23$$

Population variance;

Var(Y) = 689 x 9.23 = 6,359.47

Population standard error;

SE(Y) = 79.74 = 29.87%

95% Confidence limits;

CL = 155.66 = 58.30%

The precision of the new estimate can be expressed as:

267 ± 29.87% (95% CL 111 – 423)

Thus, an increase in area coverage by a factor of 3 gives an improvement in the precision of the estimate by a factor of about 2, but the accuracy remains the same. The conclusion is that in this particular case, aerial survey is not the appropriate technique, and alternative methodology should be considered.

2.7 Categorisation of Data

Categorisation of the quality of survey and census data is required to compare the reliability of results obtained from different techniques, under varying circumstances. Because it is important to support a continental attempt at standardisation, categorisation of data quality will to some extent follow the system used for the African Elephant Data-Base (AEDB), (Said *et al.*, 1995; Barnes *et al.*, 1999). Note, however, that the system used for the AEDB relates to a single species, with a single conservation objective.

Ideal techniques for estimating wildlife numbers do not exist. The various techniques all have their advantages and disadvantages, and are applicable in different situations, and for different purposes. In Zambia, over the past 30 years, wildlife counts have been carried out by a wide variety of donor-funded projects, by the private sector, and by the government authority responsible for managing wildlife. Firstly, to make sense of the enormous variability of counts, data quality should be classified according to specific criteria. Secondly, for future reference, standardised methods for collecting and analysing data should be used.

In the system used for the AEDB (Said *et al.*, 1995; Barnes *et al.*, 1999), data quality for individual wildlife counts is rated from 1 (best) to 3 (worst), according to different types of data collection, and count reliability from A (best) to E (worst), according to different types of techniques employed. However, some of the data quality measurements used in this system could be improved. The discussion below may seem editorial at first, but it serves one purpose, that is to show which categories of data quality are valid to use and which are not.

Direct Wildlife Counts: For sample counts or surveys, including aerial counts and line-transect foot counts, the AEDB uses sampling intensity or percentage coverage as a measure of quality, with coverage >20% rated as 1 (best). Although the AEDB has reasons for using this measure, strictly speaking this is not correct, because sampling intensity is by no means a measure of data quality. This leads to comparing apples with pears. For example, the results of an aerial sample count of elephants, carried out in a woodland area, with a 10% sampling intensity, can never be compared with the results of the same count carried out in an open area, using sampling intensity as a measure of quality. It is evident that the count done in the open area produces more accurate results than the one conducted in the woodland area. Along the same lines, two elephant surveys, both with a 10% sampling intensity, but one using photography for large herds and one without, are not comparable even though they were conducted with the same area coverage. Also surveys done with the same area coverage, but counting different species, can never be compared on account of sampling intensity. Moreover, sampling intensities greater than 15% do not improve precision, but merely increase cost (*see Chapter 6*). Frequently the suggestion that surveys with a sampling intensity of more than 20% should be rated best (Said *et al.*, 1995; Barnes *et al.*, 1999), will actually stimulate field workers to waste limited resources.

The single most important factor is the reason for counting, which determines the kind of data needed in terms of precision and accuracy. The other important factors are the species and the habitat. Therefore, categorisation should use:

1- The objective, or the reason for counting,
2- The species,
3- The habitat.

The first 3 categories merely give information on the technique and effort that should be employed under these conditions, whereas the next 4 categories provide information on the quality of the results.

4- The technique applied,
5- An indication whether the results of the methodology applied gave adequate information in relation to the objective of the count,
6- The level of precision: using the 95% confidence limits, the sample counts are presented in the following categories (Said *et al.*, 1995; Barnes *et al.*, 1999):

-**Definite** is the lower 95% confidence limit of the population estimate,
-**Probable** is the difference between the population estimate and the lower limit, and
-**Possible** is the difference between the population estimate and the upper limit.

To illustrate the above with a simple example: an aerial sample count results in a population estimate of 10,000 ± 4,000 (95% confidence limits). Hence, the lower limit is 10,000 – 4,000 is 6,000, and the upper limit is 10,000 + 4,000 is 14,000. Therefore, the number of animals in the study area is between 6,000 and 14,000. Presentation of the data is as follows: there are **Definitely** 6,000 (10,000 – 4,000) animals, **Probably** 4,000 more (10,000 – 6,000) and **Possibly** another 4,000 (14,000 – 10,000).

Data quality, in terms of precision, is rated as follows:

1. Standard error of population estimate < 10%,
2. Standard error between 10% and 20%,
3. Standard error between 20% and 30%,
4. Standard error >30%.

When relating the precision of the estimate to the objective of the count, estimates rated 1 and 2 are sufficiently precise to use for trend analysis, while estimates rated 3 and 4 may be used for regulation of management practices, such as law-enforcement effort and efficiency, or for concerns over the conservation status of particular populations.

7- The use of double counts in aerial techniques will improve accuracy of the final estimate. Hence, the last category describes accuracy by indicating whether the double-count technique was applied.

According to the system used in the AEDB (Said *et al.*, 1995; Barnes *et al.*, 1999), categorisation of aerial total counts or censuses places the results in the **Definite** group, with searching rate as the measure of quality. Searching rate can be defined as the speed of the aircraft (km/hour), multiplied by the total width of the area counted (one or two strips in km). Although it is appreciated why this measure of quality was chosen, it is not correct for the same reasons as outlined above. For example, a total count of elephants in a woodland area, using a searching rate of 200 km^2/hour, cannot be compared with a total count of elephants in wide open savanna, using the same searching rate. Within the time available to the observers, far more elephants will be spotted in the open area as compared to the woodland area. Searching patterns required by observers differ considerably for different species, while some species with large group sizes require photography. Hence, searching rate is also not a good measure of data quality when comparing counts of different species. The only solution is to use the same categories for total counts as for sample counts, including a measure of precision, obtained through repeated counts of small sections of the census area. In reality, however, most areas are only counted once, which makes it difficult to introduce this system on a continental basis.

Indirect Wildlife Counts: It is next to impossible to categorise data quality of most index techniques. For indirect techniques such as dropping counts, the 95% confidence limits of the population density, in combination with measurements of dung decay rates and defecation rates for a specific location and time of year, can be used as indicators of data quality.

Count reliability is rated from A (best) to E (worst). The rating is as follows (Said *et al.*, 1995; Barnes *et al.*, 1999):

A Total counts and counts based on individual recognition,
B Sample counts with 95% confidence limits,
C Dropping counts with 95% confidence limits,
D Sample counts and dropping counts without 95% confidence limits and informed guesses,
E Other guesses.

Distribution Data: these should be defined as **Core Range**, where individuals of a particular species are present throughout the year, **Seasonal Range**, where individuals of a particular species are present seasonally (Said *et al.*, 1995; Barnes *et al.*, 1999), and **Occasional Range**, where individuals of a particular species occur periodically.

PART II

DIRECT COUNTS

PART IIa

GROUND COUNT TECHNIQUES

Chapter 3

GROUND COUNT TECHNIQUES; AN INTRODUCTION

Ground count techniques can be divided into total counts and sample counts, carried out on foot or by vehicle. Sample counts by vehicle can be divided into terrain counts, following the same principles as sample counts on foot, and road counts, whereby the sample count is restricted to the existing road-system. The same principles apply to total ground counts on foot as to those conducted by vehicle, with the only difference that dense vegetation may not be accessible by vehicle.

3.1 Total Ground Counts

According to some of the earlier literature, total ground counts may give relatively accurate results of ungulate population abundance in areas smaller than 100 km^2 (Norton-Griffiths, 1978). The main disadvantage of total ground counts is that they are very labour intensive and therefore expensive to conduct. The advantage, however, is the simplicity of the method, both in practical terms and with regard to the theoretical concept. The census can be carried out by laymen, while the results can be worked up by anybody with a hand calculator. Unfortunately, this time-consuming and costly method is usually not as accurate as expected. Some animals will invariably be missed or double-counted during a total count. Consequently, as in sample counts, the results of a series of total counts can be very variable due to errors and biases. In total ground counts, the population size and average group size of the species under study appear to determine the variance of the individual counts (Magin, 1989). The larger the population size and average group size of the target species, the greater the variance in the individual counts (Magin, 1989). Results of 8 total ground counts of 9 ungulate species on a fenced game ranch (81 km^2) on the Athi-Kapiti plains in Kenya, gave population estimates with 95% confidence limits that averaged 46.6% (Magin, 1989). A series of 10 total ground counts of the same 9 ungulate species performed in the unfenced Nairobi National Park in 1966 (122 km^2), gave 95% confidence limits that averaged 62% (Foster and Kearney, 1967). Sample ground counts on the other hand, may give results with a better level of precision. They are also less labour intensive and therefore cheaper.

In Zambia, lechwe antelopes are often totally counted on the ground. The three subspecies of lechwe that occur in Zambia form groups of up to several hundred individuals, mostly concentrated near the edge of a water body. For a total ground count, the observers need to be in a raised position to be able to count all the individuals in large herds. Hence, the use of a vehicle is a prerequisite to obtaining adequate results.

A practical problem, however, is the abundance of water throughout the year in most of these wetland areas, limiting access to vast areas of lechwe habitat. The result is that only selected areas with proper access can be totally counted from the ground, thereby effectively changing any total count into a block sample count. In addition, as discussed above, the large average group size of lechwe will generate variable results with a low level of precision. Unless it is important for management to know how many lechwe there are in a specific isolated area with proper access, estimation of lechwe abundance should use alternative techniques. The appropriate technique may be stratified aerial sample counts, with a random or systematic survey design, using low and slow flying ultra-light aircraft, or regular aircraft, but applying photography to count accurately the individual animals in large herds.

Puku, reedbuck and oribi also follow a clumped distribution, concentrating in narrow riverine areas and on flood plains, but with small group sizes. Therefore, the abundance of these animals can be estimated accurately with total ground counts (*see section 5.6*).

3.1.1 Total Ground Counts of Hippos and Crocodiles

Total ground counts of hippos and crocodiles are a special case. Unless both species are found in swamps or other inaccessible areas, they can usually be counted from the banks of rivers, streams and lagoons. Hippos are counted during the day, whereas crocodiles are counted at night. In the Luangwa Valley, the best time of the year is the peak dry season, when most of the tributaries of the main rivers have stopped flowing and hippos and crocodiles congregate in the few remaining deep pools. Field procedures are straightforward. A small team of 3 people slowly moves along the bank, searching for animals. When the river is wide, a pair of binoculars is required. On finding a school of hippos, the team has to spend at least 20 minutes on the same spot, to make certain that submerged hippos are also counted (Figure 3.1). With crocodiles, a powerful spotlight has to be used to illuminate the eyes. In the central Luangwa Valley, with densities of 30 to 40 hippo per km of river frontage, a team of 3 observers covers approximately 25 km per day. Total cost is approximately US$ 120/day. Crocodile counts at night may be slightly more expensive, because the exercise requires two additional people with the task of protecting the team members against attacks by hippos, elephants and other creatures that are active at night. Also, crocodile counts may produce underestimates by as much as 5%, because the smallest crocodiles, below about 1 m, are very hard to spot.

Figure 3.1: Hippos are easy to count from the river bank.

3.1.2 Total Ground Counts Using Individual Recognition

Total ground counts through individual recognition are also a special case. Although this technique can be used for many different animal species, its application is especially useful for species that cannot be counted with any of the other techniques. Lion and leopard for instance, are two important species for the hunting industry, as well as photographic tourism. Their abundance is very difficult to estimate, and individual recognition through photographs and/or diagrams, indicating characteristics such as facial patterns, size and colouring of mane and other colour patterns, is a useful method for getting to know all the individuals in a population (Figure 3.2). Both lions and leopards can be baited, while lions can be called to a specific location with amplified sounds of lions devouring a prey animal. However, the method is very time consuming, and therefore relatively expensive.

Other animal species can be counted in the same way, by using specific characteristics, for instance ears and tusks in elephants, and stripe patterns in zebras (Figure 3.2).

Figure 3.2: Total counts through individual recognition use tusk size and shape, in combination with ear patterns in elephants, facial characteristics, size and colouring of mane in lions, and different stripe patterns in zebras.

3.2 Sample Ground Counts

The basic concept of a sample ground count is very simple. After defining the sample units, either quadrats, plots or strips, all individuals of the target species are counted in the defined units. Population density is then estimated by dividing the total number of animals of each target species counted by the total size of the units sampled. The density figure obtained in this way can then be applied to that part of the survey area that was not covered by the sample count. This straightforward method of linear extrapolation may produce relatively accurate results in wide open country with good visibility. In woodland areas, however, the method is usually inadequate because it is often impossible to define exactly the area surveyed, and invariably some animals will have been missed. The resulting population estimate will be an underestimate (i.e. negative bias).

In the early days of game counts on foot and by vehicle, a number of different sampling techniques was developed, using sample units of known size. The use of blocks or quadrats in ground counts is not advisable for the following reasons: blocks and quadrats return poor information on distribution, errors may be incorporated when the target species concentrates near certain features used as boundaries (rivers and streams), and uneven distribution of the target species leads to large variance between blocks and imprecise estimates. Therefore, we will limit our discussion to strips and transects as the sample units. The applicability of fixed-width strips is limited to wide open country with excellent visibility. For other situations, variable-width strips or variable visibility profiles can be used. In variable-width strips, the width of the strip is adjusted according to vegetation density (Lamprey, 1963).

For single-species counts, variable visibility profiles often produce adequate estimates of abundance in different kinds of habitat. The profile is measured by recording the distance at which all animals are seen, and then the maximum distance is used as the edge of the transect. These distances are plotted on a map, and the area of the transect can then be measured. Although they are relatively cost efficient, they have similar intrinsic problems as the fixed-width method. In multi-species counts, the larger and more conspicuous animal species will determine the transect boundary, resulting in undercounts of small species, and visibility profiles vary throughout the year, depending on vegetation density (Norton-Griffiths, 1978). For multi-species counts, they should be considered inferior to alternative, more advanced techniques. With these techniques, the average perpendicular distance is used as half the effective strip width. These ground count techniques follow the field procedures for the line-transect method. These methods can be used on foot and by vehicle, for off-road counts as well as road counts. They require recording groups of animals instead of individual animals, because the perpendicular distances are measurements taken from the transect to the geometrical centres of the groups of animals observed. These methods will be further discussed in the next chapter on line-transect counts.

Chapter 4

LINE TRANSECT COUNTS

The line-transect technique is relatively easy to apply in the field, but the underlying theory and the mathematics involved are complicated, and consequently a computer and appropriate software are essential tools. The underlying theory has been described in a series of papers. The two most important papers were a monograph by Burnham *et al.* (1980), followed by the volume by Buckland *et al.* (1993), which documents the most recent developments in the science of line-transect methodology.

4.1 Theoretical Background

The line-transect method depends upon the fact that the probability of detecting (p) an animal or a group of animals from the line of travel or transect declines with the animal's distance from the transect. An animal directly on the transect has a probability of detection of 1 ($p = 1$), but an animal far from the transect is less likely to be spotted by an observer ($p < 1$). Thus, the probability of detection p ranges from 0 (animal not visible from the transect) to 1 (animal on or very near the transect).

In line-transect sampling the observer progresses through the study area following a straight line of known length, the transect. The observer records each group of the target species and notes the distance of the group from himself (*sighting distance*). The sighting distance can be measured with a range-finder, a measuring tape or by pacing. He uses a compass to note the bearing of the group; he will use this to calculate the *sighting angle* relative to the transect (Figure 4.1). Using the sighting angle and the sighting distance, the perpendicular distance can be calculated for each group of animals. Thus, diagrammatically a line-transect sample consists of a number of perpendicular distances representing the distances from the transect line to the geometrical centres of the groups of animals (Figure 4.1).

A frequency diagram of grouped perpendicular distances will show the probability of detecting a group of animals at particular distances from the transect line. Mathematically, this relationship is represented by a curve called the detection function g(x), where x is the perpendicular distance (Figure 4.2). The detection function is the conditional probability of observing a group of animals, given that the group is at a particular perpendicular distance from the transect line.

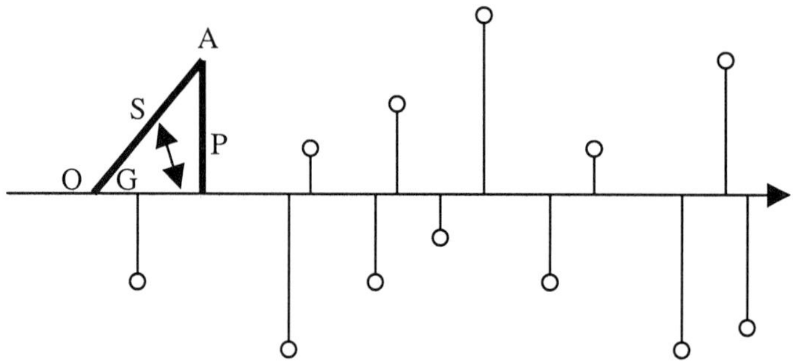

Figure 4.1: Diagrammatic representation of a line-transect sample, where the centre-line is the transect line, with 13 observations of the target species, indicated by perpendicular distances. The thick lines show the data that should be collected for a single observation, where: O is the position of the observer, A is the position of the group of animals, S is the sighting distance and G is the sighting angle (P is the perpendicular distance).

The probability of detection is not only a function of the animal's distance from the transect and the density of the vegetation, but is also related to such factors as the conspicuousness of the species (size, colour pattern and behavioural traits such as flight distance), the timing of the survey, and the average group size. To a lesser extent it is also related to the alertness, interest and training of the observers. For instance, in open woodland, a large group of impala can be detected at a much larger distance than a single bushbuck, whereas the curiosity of the wildebeest makes them much easier to detect than the shy eland. Also, in areas where illegal hunting is prevalent, the flight distance of most animals will be greater than those inhabiting an area without hunting. Therefore, for a particular area, each animal species will show different detection probability characteristics. Similarly, the detection probability curve for the same species will vary between areas with different habitat and hunting conditions.

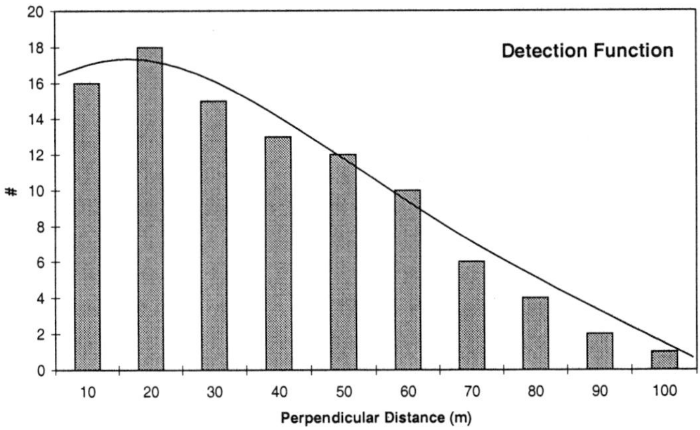

Figure 4.2: Hypothetical example of a frequency diagram of perpendicular distance data for groups of animals observed (#). The detection function g(x) is indicated.

4.1.1 Conceptual Background

Now we have reached the difficult part of line-transect theory, which is to use the set of perpendicular distance data summarised in the frequency diagram (Figure 4.2) to model the detection function g(x). We need to calculate the detection function in order to estimate the density of the target species. The shape of the detection function g(x) determines the type of estimator(s) that can be used to describe the curve in mathematical terms.

Although the mathematics required to model the detection function g(x) are rather complicated, the conceptual background of line-transect estimation is quite simple. Any estimator of the density (D) of (n) observations of animals or groups of animals over a transect line with length (L) can be expressed as:

$$D = \frac{n}{2a \times L} \text{ or } \frac{n}{2L} \times \frac{1}{a} \qquad (1)$$

The unknown parameter (a) can be roughly described as one-half the effective transect width, where the relationship between (a) and the detection function g(x) is the integral of the detection curve:

$$a = \int_{0}^{w} g(x)d(x) \quad (2)$$

Thus, the main problem in line-transect estimation of density is to estimate the parameter (a) or its reciprocal 1/a. The probability density function of the perpendicular distance data f(x) is related to g(x) as: f(x) = g(x)/a, where both f(x) and g(x) have exactly the same shape (Burnham *et al.*, 1980). Because animals on the transect line should always be seen with certainty, g(0) = 1 and therefore f(0) = 1/a. Thus, equation (1) is often expressed in terms of f(o):

$$D = \frac{n \times f(0)}{2L} \quad (3)$$

The problem has now been simplified to estimating f(0), of which the units are the reciprocal of the perpendicular distance (1/x). An example using a fixed-width strip of width 2w and length L will clarify this (Burnham *et al.*, 1980). Assuming that no animals within the strip are missed, the detection function g(x) = 1 and the probability density function f(x) = 1/w. Thus f(0) = 1/w and the density estimate is:

$$D = \frac{n}{2L} \times \frac{1}{w}$$

In other words, fixed-width strips are a special case of line-transects.

4.1.2 Different Shapes of the Detection Function

There are many possible shapes of the detection function. Some of the estimators used for modelling the detection function g(x) have a simple estimator of f(0). For example, f(0) is 1/a for the Negative Exponential estimator, and f(0) is 1/w for the Uniform estimator, which is similar to using the mean perpendicular distance as half the effective transect width (w). Others, however, are more complicated than this.

The results of repeated line-transect counts in the Lupande GMA in Zambia show us that the underlying detection function g(x) has one general shape, and two less common ones.

The most common shape is the frequency diagram with an underlying negative exponential or reverse logistic detection function (Figure 4.3, Box 4.1). Most of the data sets from the Lupande GMA were from these detection functions. Three different estimators are suited to model these shapes, e.g. Negative Exponential, Exponential Polynomial and the Exponential Power Series. The Negative Exponential estimator frequently provided the best fit for data from 10 ungulate species collected in the Lupande GMA between 1992 and 1997. However, as line-transect theory evolved we began to analyse the same perpendicular distance data with the Hazard estimator with Cosine or Polynomial adjustment terms (Box 4.1). This estimator sometimes provides a good fit for the detection function g(x).

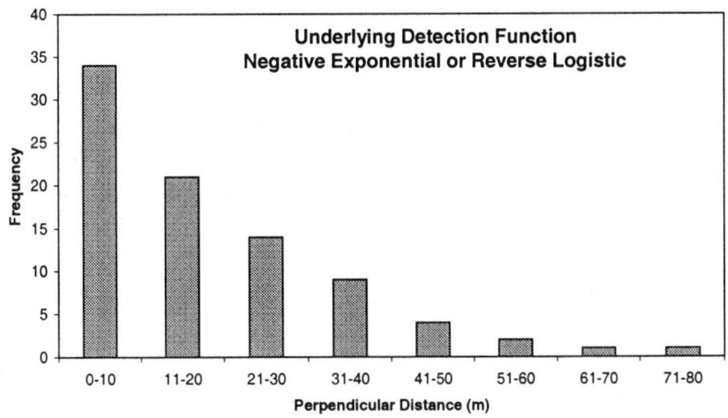

Figure 4.3: Hypothetical example of a frequency diagram of perpendicular distance data from an underlying negative exponential or reverse logistic detection function g(x).

The second shape of the detection function g(x) from the Lupande GMA is that from an underlying uniform detection function (Figure 4.4, Box 4.1). The uniform detection function is a straight line that runs parallel to the abcissa (g (x) = $^{1}/_{w}$). Some small data sets from the Lupande GMA were from this detection function. However, when the sample sizes were increased, these data sets usually appeared to come from an underlying negative exponential detection function. Thus, not only the type of data, but also the sample size determine the shape of the detection function. Data with an underlying uniform distribution can be analysed with the uniform estimator. A linearly declining detection function can be analysed with the Uniform estimator with Cosine adjustment terms (Fourier Series).

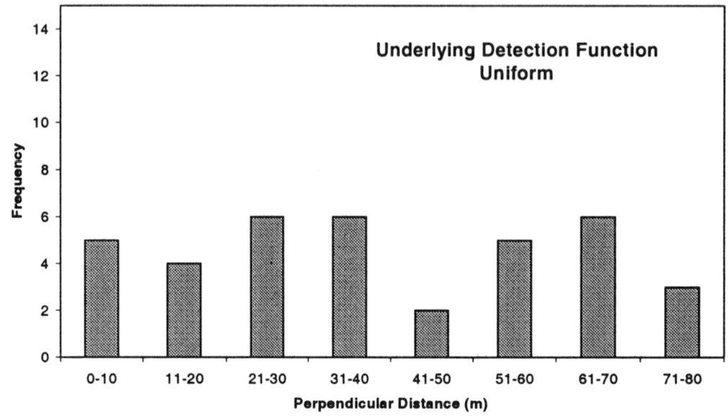

Figure 4.4: Hypothetical example of a frequency diagram of perpendicular distance data from an underlying uniform detection function g(x).

Only a few data sets from Lupande GMA were from the third shape. This one has an underlying half normal detection function (Figure 4.5, Box 4.1). Here, the best fit would be obtained by using either the Half Normal estimator or the Fourier Series estimator to model the detection function g(x).

Box 4.1: Formulae for the different shapes of the detection function (g(x)), the Hazard estimator and the Uniform estimator with a single Cosine adjustment term.

Negative Exponential:	$g(x) = e^{-ax}$
Reverse Logistic:	$g(x) = \{(1+b)e^{-ax}\}/\{1 + be^{-ax}\}$
Uniform:	$g(x) = 1/w$
Half Normal:	$g(x) = e^{(-x^2/2a_1^2)}$
Hazard estimator:	$g(x) = 1 - e^{-(x/a_1)^{-a_2}}$
Uniform plus Cosine adjustment term:	$f(x) = 1/w(1 + a_1\cos(\pi x/w))$

When reading this section, however, we must keep one thing in mind: although the Lupande GMA is considered one of Zambia's prime hunting areas, wildlife densities are too low to provide large sample sizes with the budget that is available. The survey area was always covered three times a year, at a cost of approximately US$ 7,500 per year. Between 60 and 80 sightings are recommended by Buckland *et al.* (1993), but our annual sample sizes are much lower than this. To obtain an appropriate number of sightings for most target species counted in the Lupande GMA would require an annual budget for line-transect surveys of at least US$ 15,000, or about US$9/km² (the survey block is 1,660 km²). This is where line-transect methodology, or any other type of transect sampling on foot, becomes a problem when counting large African ungulates. The methodology is extremely useful and the best option available when assessing high densities of particular species in the developed world, where money is usually not a limiting factor. Also, the methodology is very useful for indirect methods such as counting droppings that are stationary, accumulate over long periods and are therefore usually available in large quantities. However, budget constraints and low densities of the target species sometimes limit the use of this technique. Many conservation areas on the continent have lower densities of wildlife than the Lupande GMA, and certainly lower budgets for this type of exercise.

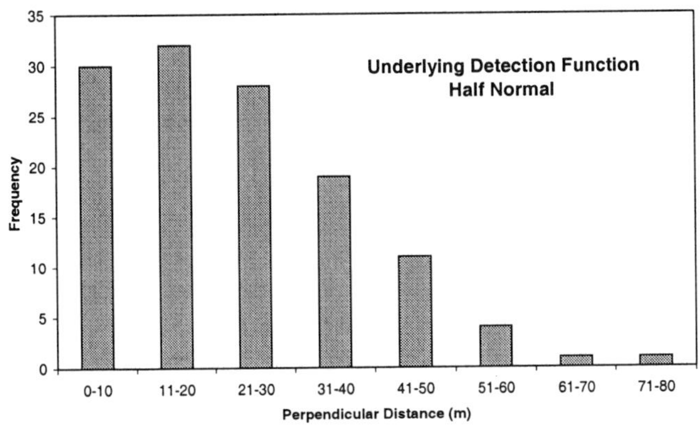

Figure 4.5: Hypothetical example of a frequency diagram of perpendicular distance data from an underlying half normal detection function g(x).

The most common shapes of the detection function from the Lupande GMA are illustrated by data from 6 large ungulate species. The data were collected during three line-transect surveys in 1996 and three in 1997 (Figure 4.6). From 1996 to 1997, animal densities in the Lupande GMA were assumed to have remained the same. The data for 1996 and 1997 were combined to increase the number of sightings: Grysbok (44),

Common Duiker (75), Warthog (77), Impala (113), Bushbuck (27) and Elephant (35).

In all cases a negative logarithmic curve provided the best fit. A negative exponential curve gave the second-best fit. For analysis of abundance, the Negative Exponential or the Hazard estimator with either Cosine or Polynomial adjustment terms provided the best fit in all cases.

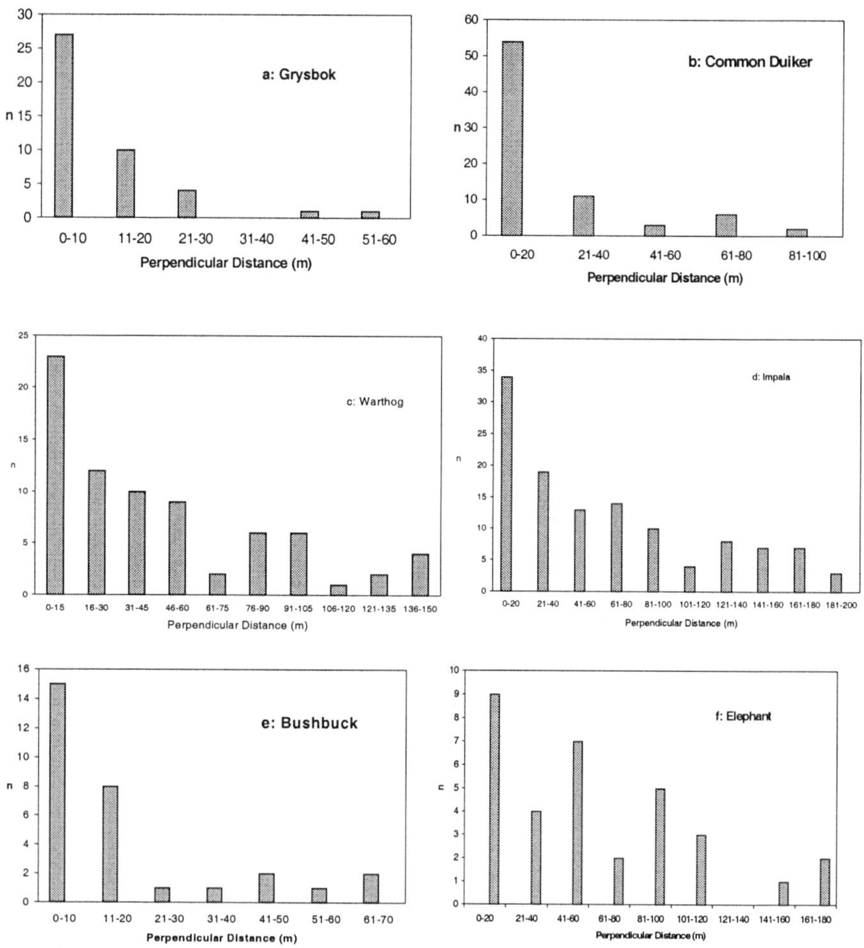

Figure 4.6: Frequency diagrams of perpendicular distance data for 6 large ungulate species collected in the Lupande GMA in 1996 and 1997. a) Grysbok, b) Common Duiker, c) Warthog, d) Impala, e) Bushbuck, and f) Elephant. With the exception of bushbuck, 1996 and 1997 data are combined. Outliers were omitted.

4.2 Software to Analyse Line Transect Data

There are four software packages available for analysing line-transect data: the first program, TRANSECT, was developed by Burnham *et al.* (1980), and uses 5 different estimators to model the detection function g(x), e.g. Fourier Series, Exponential Power Series, Exponential Polynomial, Negative Exponential and Half-Normal. The program is not very user-friendly and requires advanced understanding of line-transect theory, mathematics and rudiments of using a computer. The second program, DISTANCE, is designed to be used in conjunction with Buckland *et al.* (1993) and replaced TRANSECT. DISTANCE differs from its predecessor in its advanced analysis methods and expanded capabilities. The program uses 4 general estimators to model the detection function g(x), e.g. Hazard, Uniform, Half-Normal and Negative Exponential, each with Cosine, Hermite and Polynomial adjustment terms. The program can select the best model using all possible combinations of estimators and adjustment terms. For advanced users who are familiar with the program's command language and line-transect mathematics, a wide range of options allows comprehensive analysis of line-transect data. But it is not particularly user-friendly and requires a solid understanding of line-transect mathematics. The good news is that in 2001, a Windows version of DISTANCE will come on the market. A new program called LOPES will also soon come on the market, which is a user-friendly version of DISTANCE. The fourth program, ELEPHANT by Dawson and Dekker (1992), is user-friendly. It is based on Burnham *et al.* (1980), and uses the Fourier Series estimator to model the detection function. The main problem with program ELEPHANT is that it uses only the Fourier Series as an estimator. The Fourier Series estimator assumes the perpendicular distance data follow an underlying half normal detection function g(x). Figure 4.6 shows that with small sample sizes (few sightings) most perpendicular distance data derived from large African ungulates have an underlying reverse logistic or negative exponential detection function g(x). The Fourier Series was at first considered a robust estimator that fitted a variety of situations (Burnham *et al.*, 1980). However, as line-transect theory evolved, statisticians concluded that other models, like the Hazard rate, were much better. When using DISTANCE or LOPES to analyse population abundance, the Hazard or Negative Exponential models, using Cosine or Polynomial adjustment terms, usually provide the best fit for large African ungulates.

4.3 Survey Design

As mentioned before, one of the major advantages of the line-transect method is the relative ease of its implementation in the field. However, prior to any survey, it is of the utmost importance to consider carefully the following factors:

1. *The objective*: The reason for the count is the first consideration when designing a survey. Determine the desired level of sampling effort and its allocation over both the study area and the duration of the study. Factors that govern the level of effort include the number of transect lines, the length of each line, the frequency with which each line is surveyed, and the number of teams of observers. The effort depends upon the accuracy and precision required by the objective of the study. If the objective is to collect data for quota-setting purposes or population reduction by culling, we need accurate estimates. Unfortunately, we will almost never know whether an estimate is accurate or not, and we will therefore have to limit sources of bias when designing the survey. Analysis of population trends requires precise estimates. This implies a large number of transects, repeated coverage and possibly a stratified survey design to limit variation in the final results. If the objective is to collect general information on animal abundance, to be used as feedback for law-enforcement operations for instance, targets are set in terms of 'limits to acceptable change'. Then, population estimates do not need the level of precision that is required for population trends, reducing the cost of the exercise. However, regular surveys with the same objective should be carried out with a standardised design to avoid confounding differences in the final results.

2. *Logistics*: Determine the number of staff and vehicles available for the exercise. These determine the size of the area that can be covered in a single day and the time needed to finish the study. Large areas covered quickly will limit bias due to movement of animals between transects. Repeated coverage of the same transects within the same time of the year increases the number of sightings, and therefore the precision of the estimates. Logistics are closely related to the budget, which in turn determines whether the objective can be achieved.

3. *Budget*: Determine the exact size of the study area and the extent of the road-system. These, in combination with the objective of the study, determine the level of effort and therefore the size of the budget required for the exercise. A study area with an extensive road-system facilitates the transportation of teams of observers to the start of the transects and collection at the finish. A study area that lacks infrastructure will therefore require a larger budget than one with good access. Also, a good road-system will facilitate the establishment of permanent transect lines.

4. *Placement of transects*: Determine the placement and spacing of transect lines. Transect lines may be either temporary or permanent. Permanent transect lines should be considered if the transects are to be surveyed periodically. Permanent transects enable pooling the data, thereby improving the precision of the final population estimates. Permanent transect lines can be delineated by markers, or alternatively, one can find the transect starting point with a Global Positioning System (GPS). Transect lines can be placed at random or systematically. If the objective includes analysis of spatial distribution of the target species, transect lines should be placed systematically. The lay-out and spacing of transects will vary according to statistical design requirements, ecological factors, considerations of logistics, supply and access, and the objective. If there is no information on wildlife abundance from previous surveys, a pilot survey may be required to get a general idea of animal densities.

5. *Timing*: The number of animals seen will vary according to time of day and season (7 to 10 AM during the peak dry season are the best times).

4.3.1 Stratification of the Survey Area

Stratifying the survey area by features such as habitat type or differences in the density of the target species is sometimes helpful in reducing confidence limits. In most cases, however, it is not advised because a stratified design presents a potential pitfall. The strata with low densities of the target species must contain a sufficient length of transect line to provide enough sightings to give an estimate for each stratum. Experience shows that in most low-density strata there are few sightings of animals. Therefore a large number of transects would be required to obtain enough sightings to justify analysis, thereby increasing the cost of the survey.

To improve the precision of the final estimate by stratification, prior information should be available on the distribution of the target species and the densities in each stratum. Also, the density differences between strata must be high, if there is to be a significant improvement in precision. When several species are counted during the same survey, as is usually the case, a single stratification of the survey area, accounting for differences in the densities of all species, will be difficult to find. In summary, stratification is a useful tool to improve the precision of the estimate when only one species is counted, when information is available on the distribution of that species, and when there are significant density differences between strata.

To illustrate stratification, we will use an example from the Lower Lupande GMA. This GMA can be divided into 6 landscape units, each with a different topography, vegetation and wildlife component (Figure 4.7). The first unit is the alluvial valley floor, between the river and the escarpment foot hills, with rich alluvial clays and vegetation dominated by mopane and riverine woodland. Most of the wildlife in the Lupande area is found here. Units 2, 3 and 4, the foot hills, escarpment and reverse slope respectively, are part of the escarpment block, with shallow, nutrient-poor surface soils, and predominantly a scrub-type vegetation. Throughout most of the dry season, these three units lack water, and, since food quality is poor there is little to no wildlife. Unit 5, lowland mixed woodlands with transitional mopane/miombo woodland and thickets, supports low densities of small antelopes, some kudu and Lichtenstein's hartebeest, in addition to the occasional giraffe and elephant.

The main objective for conducting line-transect counts in the Lower Lupande is to provide management with estimates for quota-setting purposes. Most of the hunting takes place in landscape unit 1, the alluvial valley trough, where most of the wildlife concentrates. The survey design in the Lower Lupande eliminates landscape units 2,3,4 and 6 (Figure 4.7). The alluvium (unit 1) is taken as stratum 1, and sampled at a high intensity, whereas the lowland area (unit 5) is sampled at a low intensity, mainly to provide management with *indicators* for adjustment of annual off-take quotas for both kudu and hartebeest.

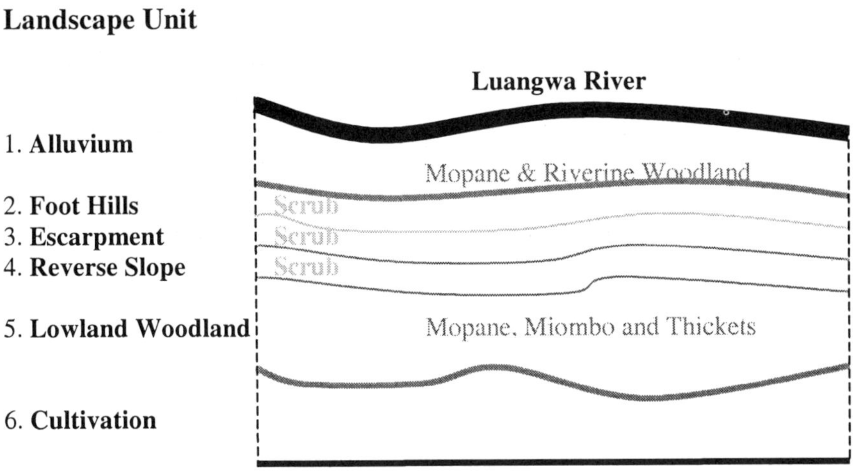

Figure 4.7: Diagram of broad landscape units found in the Lower Lupande Game Management Area in the central Luangwa Valley, Zambia.

4.4 Field Procedures

The following assumptions are critical to obtaining reliable estimates of population abundance from line-transect sampling.

- Defining a straight line of travel using a compass or a GPS;
- Ensuring that all animals on and very near the line of travel are detected with certainty (probability of detection is 1).
- Obtaining accurate measurements of sighting distances, using a range-finder or the method of pacing, and accurate measurements of angles (sighting angle), using a compass;
- Ensuring that all measurements are taken from the transect line to the geometrical centre of the group of animals where they were first spotted. Thus animals are fixed at the initial sighting position.
- Obtaining accurate measurements of transect length, using measurements from a map, a GPS, pacing or a measuring wheel or hip chain (topofil).

In the field, a team ideally of 3 observers progresses through the area following a straight line of known length (transect line). The recorder controls the line of travel, takes measurements and keeps notes while the other two are observers. The recorder takes notes of each animal or group of animals observed and measures the distance of the animal(s) from where he is standing. He uses a compass to note the bearing which can later be converted to a sighting angle relative to the transect line. Note that sighting angles should never exceed 90°. In the case where a sighting angle exceeds 90°, the complementary angle has to be used; e.g. the sighting angle is deducted from 180° to give the complementary angle. The recorder can then calculate the perpendicular distance of each animal or group of animals from the transect line, using the following formula (Figure 4.8):

$$\text{Sin}(S_a) \times D_s = D_p$$

Where:

S_a = Sighting Angle,
D_s = Sighting Distance,
D_p = Perpendicular Distance.

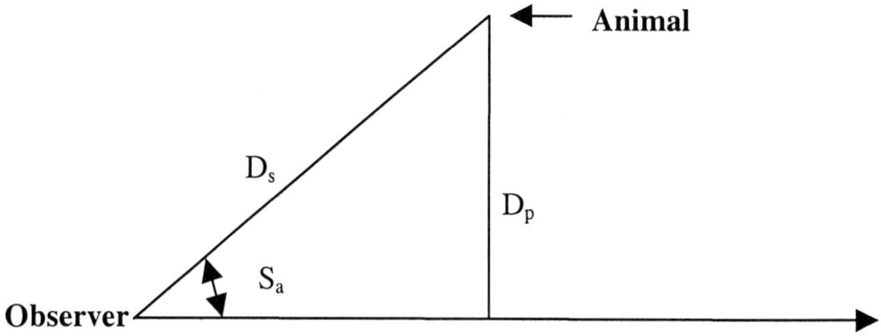

Figure 4.8: Diagrammatic representation of the measurements required to calculate the perpendicular distance of an animal from the transect line.

We can treat study areas with a well-established and extensive road-system differently from those with perimeter roads only, or no roads at all. In a protected area, the roads are usually built in scenic places with high densities of wildlife, such as alongside rivers and streams, along contours and through valleys. Wildlife densities usually follow a declining gradient away from the river, stream or valley. To minimise bias and variation between transects, transects should be laid perpendicular to rivers, streams or any other features attracting high densities of wildlife. Thus, for study areas with a road-system that was built for game viewing or hunting, the transect lines should cut the roads at a 90° angle. Each transect should start and finish at a road. The length of each transect should be such that it can be covered by a team of observers during a single morning. The exact length of each transect can be measured from the map. When a proper map is not available, the length of each transect can be assessed with the use of a GPS, by pacing, or using a measuring wheel or hip chain (topofil).

For areas without a road-system, the field procedure is slightly more complicated. The first problem is the difficulty of estimating accurately the length of each transect line. Precious time may be wasted when the team has to return to a pick-up point such as a perimeter road. When insufficient GPSs are available, the most practical solution is to measure transect length and sighting distance by pacing. Each recorder uses his or her personal step-length, which is determined prior to the survey, to step out the transect line and the sighting distances, using a pre-printed step-length sheet, ticking off every hundred steps. Prior to the survey, each recorder has to cover a distance of exactly 100 m ten times, to accurately determine the personal step-length. In the Lupande GMA, in 1996, the individual step-length of 16 different recorders varied from 72.3 to 92.5 cm, with a mean standard deviation of only 1.83% (Jachmann, 1996a). After the survey, all measurements of transect length and sighting distance will have to be converted to actual distances by using the personal step-length of each recorder. Pacing enables the transect length to be determined with an accuracy greater than 98% (checked with a GPS). However, in the field a team of observers frequently has to avoid obstacles such as trees and rock formations, resulting in an effective transect length that is less than the number of paces counted. Repeated surveys in the Lupande GMA showed that on average an additional 2.5% was covered on each transect due to obstacles on the line of travel, giving a total error of 4.5% (2% + 2.5%). Obviously this applies to the Lupande GMA, and elsewhere this correction factor can be checked with a GPS.

The second problem in areas without roads results from teams of observers returning to the beginning of the transect without collecting observations. This, however, can be avoided by using block transects instead of line transects. In block transects, a team of observers first covers a transect away from the drop-off point, then walks at right angles, still collecting data. Finally they turn again and walk back to the road, continuing to collect data (Figure 4.9). Note that this is only valid (i.e. the section across) when a density gradient of species counted does not exist. This was the case in the Lupande GMA for the first 7 km away from the Luangwa River.

The compass is an important instrument for maintaining a straight line on the correct bearing, as well as for measuring the bearings of animals or groups. A compass is relatively simple to use and all recorders participating in a survey should be proficient with the instrument. The compass needle will always point to the magnetic north, which is slightly to the west or to the east of the true north. The difference between the magnetic north and the true north is called magnetic variation and depends on the position of the observer on the planet earth (Figure 4.10). For instance, in the Lupande GMA, the variation is approximately 4° west, implying that each compass bearing will have to be increased by 4° to give the bearing to the true north. However, even without correcting for magnetic variation, transect lines will run parallel.

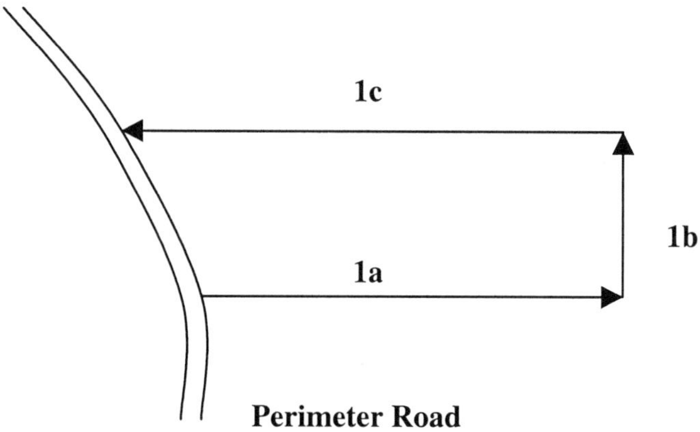

Figure 4.9: Diagram of a block transect starting at a perimeter road and consisting of 3 segments. Segment 1a is away from the road, 1b is across and 1c is back to the road.

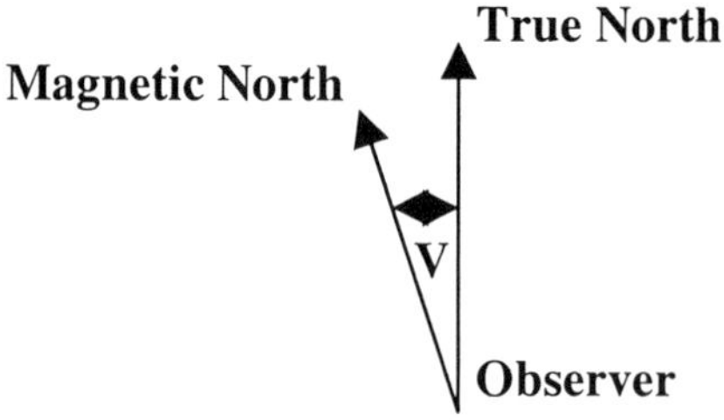

Figure 4.10: Diagram illustrating magnetic variation (V).

To follow a straight line in the field, set the appropriate bearing on the compass ring, point the sight of the compass to an object as far away as possible (such as a tree), and walk towards that object from where the next reading is taken. When observing an animal or group of animals, point the sight of the compass to the animal(s) or the point at which the animals were first seen, and take the reading while standing on the transect line. Always make sure that all measurements are taken from the transect line. When measuring the sighting distance of a group of animals, make sure that the measurement is taken from the transect line to the geometrical centre of the group of animals. The observer's feet should never leave the line of travel.

Last, but certainly not least, all data collected in the field should be recorded on a data sheet, with observations on the front and distances in the form of steps on the back. For each observation, note the species, group size, take the bearing relative to the line of travel (sighting bearing) and step out the distance to the location where the animal(s) were first seen (sighting distance), and finally note the number of paces covered on the transect. An example of the data sheet used for block surveys (e.g. Figure 4.9) in the Lupande GMA is provided in Table 4.1.

Table 4.1: Example of the data sheet used for block transect surveys in the Lupande GMA.

BLOCK TRANSECT SURVEY DATA SHEET				
Date:…………………….. Survey Block:………… Transect Number:……..				
	# Segment	**Compass Bearing**	**# Steps**	**Recorder Team Leader**
Away from Road				
Across				
Back to Road				
Species	**Group size**	**Sighting Distance (Steps)**	**Sighting Bearing (^{0})**	**Remarks (Steps)**

4.5 Precision and Accuracy of Line Transect Results

The precision of a population estimate depends on the abundance of the target species, its spatial distribution over the study area and the number of transects covered during the survey (sample size). Because analysis deals with sightings of groups of animals and mean group size, the number of sightings of groups provides a better measure to predict the precision of the resulting estimate than absolute numbers or density (Burnham *et al.*, 1980; Buckland *et al.*, 1993; Jachmann, 1996b). Generally, an even distribution with small group sizes will give a more precise estimate of abundance than a clumped distribution with large group sizes. Also, a large number of transect lines (sample size), whether as part of the initial lay-out or as repeated surveys of the same transects, will result in more precise estimates of abundance.

The line-transect data collected in the Lupande GMA in 1996 and 1997 illustrates the relationships between precision and numbers, precision and density, and precision and number of sightings. The survey was conducted in a study area of 1,660 km². Forty-one transects were placed at 5 km intervals, giving a total length of approximately 318 km. Each year, the transects were covered three times during October, effectively increasing the number of transects to 123 and the total length to 954 km per annum (Jachmann, 1996a). Note that pooling of the data was only valid because there were no changes in animal density during this one-year period.

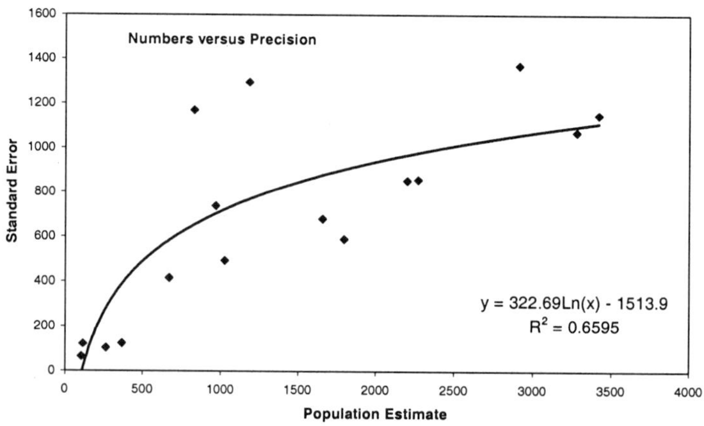

Figure 4.11: Standard error in relation to population estimate for 10 large ungulate species counted during six line-transect surveys in the Lupande GMA in 1996 and 1997.

The detection function was modelled with the Hazard, Negative Exponential and Uniform estimators. The relationship between the standard error and numbers of 10 large ungulate species follows a logarithmic curve. In other words, higher numbers of animals result in larger standard errors (Figure 4.11). High animal densities result in large group sizes and consequently in clumping of the population and estimates that are less precise (*see section* 6.4 *for a more detailed explanation*). In relative terms, however, the % standard error declines with increasing densities (Figure 4.12). Although the relationship is not significant, at densities of up to approximately 0.7 animals per km² there is a quick initial increase in precision, with a gradual improvement in precision at higher densities (Figure 4.12). The number of sightings of a particular animal species is closely related to the sample size, but evidently also to density.

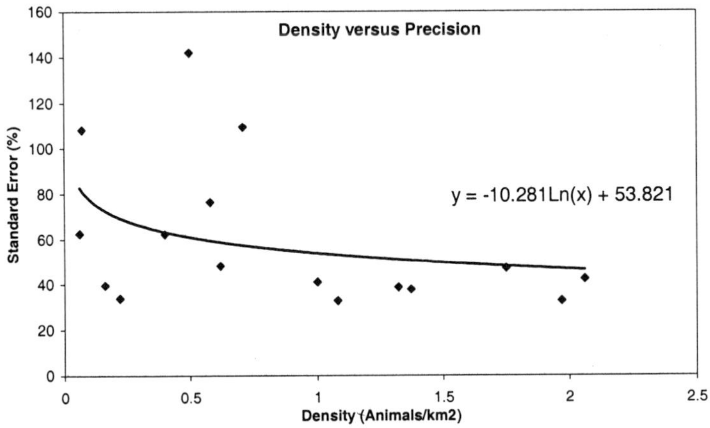

Figure 4.12: The precision of the population estimate (% Standard Error) in relation to the density of 10 large ungulate species, counted during six line-transect surveys in the Lupande GMA in 1996 and 1997.

As one increases the sample size, there is a quick initial increase in precision up to approximately 15 sightings, after which a further increase in the number of sightings results in a gradual improvement in precision (Figure 4.13). The number of sightings is a better prediction of the precision of the final estimate than the absolute density.

With the equation in Figure 4.13 we can calculate that a sample of 60 sightings will result in an estimate with a standard error of approximately 14%, and a sample of 80 sightings in a standard error of about 5%. However, this is based on small samples collected in the Luapande GMA.

Larger samples, or data collected elsewhere, may give different results. Also, there will be less spread of the data, when modelled for a single species. The practical implications are that a line-transect survey should be designed in such a way that a minimum of 60-80 sightings are collected of each target species counted, (Buckland *et al.*,1993).

Permanent transect lines should be established to enable pooling the data of repeated surveys to increase the number of sightings. Transect lines should be placed systematically if one of the objectives is to generate information on the spatial distribution of the target species. Transect lines should not be too close, so as to avoid bias through movements of animals from one to the next transect line. If the spacing of the transect lines is too wide, and therefore the number of sightings too low, they can always be covered again. In practice, however, the statistical design requirements discussed above are usually outweighed by one single factor, which is the budget.

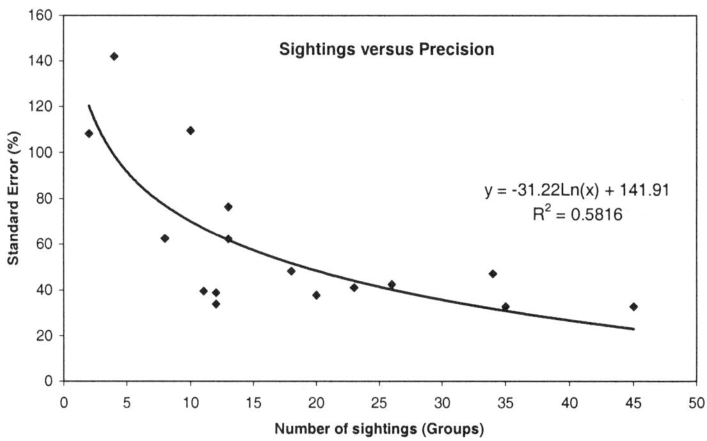

Figure 4.13: The precision of the population estimate (% Standard Error) in relation to the number of sightings of 10 large ungulate species, counted during six line-transect surveys in the Lupande GMA in 1996 and 1997.

We would like every survey to yield estimates that are both precise and accurate. Unfortunately, we will almost never know the accuracy of an estimate, and we certainly do not have a measure to predict it. Depending on the species and its distribution over the study area, we may need a much smaller sample for an accurate estimate than for an estimate with an acceptable level of precision. However, to obtain an accurate estimate, our only option is to limit potential sources of bias. With line transects, the following sources of bias may affect the accuracy of the final results:

- The probability of detection on or very near the transect line should be one. This assumption does not always hold, and depends on the species, the habitat and the number and speed of the observers. The number of observers should be no less than three, walking at a slow constant pace, keeping noise levels low.

> This is also one of the reasons why line-transect methodology should not be used in aerial surveys. First, the ground under the aircraft cannot be searched, second, an animal under a tree may be missed close to the line of travel, and third, an observer cannot search all parts of the strip at once.

- Animals should not move in response to the observers. Density estimates will be biased upward for movement towards the observers, and biased downward for movement away from the observers. Measurements (distance and bearing) should be made from the observer to the geometrical centre of the group of animals where they were first spotted.

- The probability of sighting a group of animals depends not only on its distance from the transect, but it may also depend on group size. This is known as size bias, which will be tested and corrected when using the appropriate software, such as program DISTANCE.

- Movement of animals between adjacent transects that are counted simultaneously may result in overestimates. This bias can be avoided by spacing transects not too close together, and by synchronising starting times for teams of observers counting adjacent transects.

4.6 Data Analysis with Program DISTANCE

The data from a line or block-transect survey are a set of distances and angles, the number of groups observed, and the number of transects covered. Each sighting distance and sighting angle is transformed to a perpendicular distance for that group of animals. The perpendicular distances are then used to calculate the density of the target species in the study area. Discussion of the underlying mathematics of the estimators and the adjustment terms is beyond the scope of this book. For a detailed discussion, see Buckland *et al.* (1993).

DISTANCE works with a command language, where the user tells the program which steps to execute by entering commands and data. The program can be used either in the interactive or batch mode. With small sample sizes, the interactive mode is preferred, and the data are entered from the keyboard. With large sample sizes it is advisable to use the batch mode where commands and data are stored in files created by the DOS Editor. The program requires three types of information to execute an analysis: OPTIONS defines the type of sampling, data and measurements used; DATA defines the sampling structure, sampling effort and observation data, whereas ESTIMATE defines the type of analysis performed.

Using the DOS Editor, the output can be read and printed from the Dist.out file. The type of output will vary depending on the analysis. The general structure is as follows:

- A listing of estimation options and a glossary of terms.

- The probability function estimation, including model fitting and selection, parameter estimation table, detection probability plots (frequency diagrams), and Chi-square goodness-of-fit tests.

- Expected cluster size estimation, which is generated only if group sizes are used (Object = Cluster in OPTIONS).

- Density estimation results, which is a summary table of estimation results with density, and abundance if the size of the survey area is indicated.

- Summary, showing the statistics for all models (Pick=None in ESTIMATE), or the model with the best fit, including a density estimate, numbers of animals in the survey area and cluster analysis on mean group size.

If specified, the program selects the best model based on the chosen decision criterion, which is either minimising *Akaike's Information Criterion* (*AIC*), or a *likelihood ratio test*, which compares the computed probability to the specified significance level (Pvalue in OPTIONS). Both criteria will almost always select the same model. The output will include messages about problems encountered with fitting a particular model, whereby the most important message is that key parameters are at upper or lower bound, suggesting that the user requested an unreasonable fit and may need to truncate the data or use different classes for the frequency diagram. For each model, a plot of the detection probability function is provided, followed by a Chi-square goodness-of-fit test. If $P < 0.05$, the estimated function deviates too much from the frequency diagram of perpendicular distance data. In this case the user may change the intervals used for constructing the frequency diagram and Chi-square test with the GOF command in ESTIMATE. For ungrouped data, without a GOF command, 3 default sets of intervals are used ($\sqrt{n}$, $0.67\sqrt{n}$ and $1.50\sqrt{n}$, whereby n is the number of observations).

Because the observed average cluster size (group size) may change with increasing detection distances, a size-bias regression is performed between detection probability g(x) and observed cluster size. If the slope is not significantly different from zero, there is no significant size-bias, and the expected cluster size is estimated by the average cluster size. Without significant size-bias, the average cluster size will be more precise.

For detailed information on the use of the program you will have to consult the user's guide (Laake *et al.*, 1994).

4.6.1 Analysing the Lupande Warthog Data with Program DISTANCE

The following example uses the warthog data from the Lupande GMA, collected in 1996 and 1997. The data for both years are combined to give a sample size of 77 observations of groups of warthog. Warthog densities were the same for both years. Twenty block transects, each consisting of three segments, were counted six times in October. For the purpose of this example, each block transect was considered as a single transect, while the 6 repeated counts of each block transect were pooled to simplify analysis and to improve the precision of the estimate. The first step is to visually examine the data, by constructing a frequency diagram with the perpendicular distance classes on the abscissa and the number of groups observed in each class on the ordinate. As a rule of thumb, the number of classes should be approximately $\sqrt{n}$. Thus, in our example we should have approximately $\sqrt{77} = 8.8 = 9$ classes. The perpendicular distance data in our warthog example range from 0 m to 262 m. The most convenient number of classes is 9 of 30 m each (Figure 4.14).

Inspection of the frequency diagram shows us that the perpendicular distance data most likely come from an underlying negative exponential detection function. In program DISTANCE, the default sets of intervals used for the frequency diagram are 6, 9 and 13 (see above). Either the Negative Exponential estimator or the Hazard estimator will be the correct choice for analysing the data.

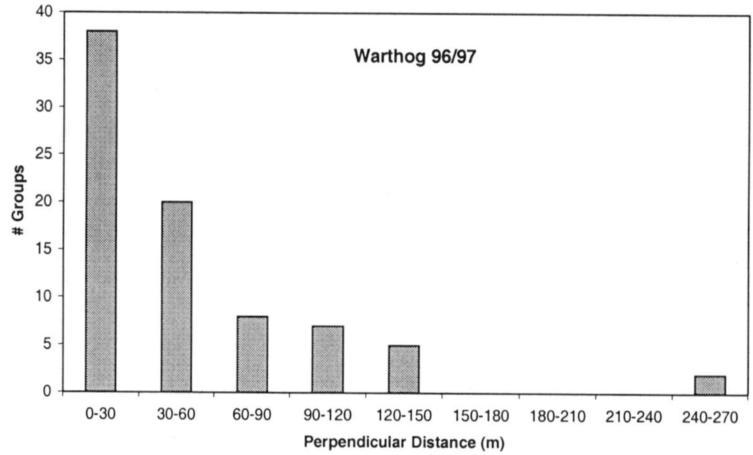

Figure 4.14: The perpendicular distance data for warthog, including two outliers, obtained through a series of 6 line-transect surveys in the Lupande GMA in 1996 and 1997. The data probably come from an underlying negative exponential detection function.

The program allows a maximum of 5 different combinations of estimators with adjustment terms per run. First, the data will be analysed with the Hazard, Negative Exponential, Uniform and Half Normal estimators, using Cosine, Hermite and Polynomial adjustment terms. This will require 3 runs, with the best model from the first run included in the second run and the best model from the second run included in the third run. Then, the same procedure will be performed without the two outliers. This will require an additional 3 runs. The program will select the models with the best fit.

The next step is to create a data file (Dist.dat for instance) in the DOS Editor, which requires only three things:

1. To indicate the length of each transect line in km using sample/effort=(?),
2. To indicate the perpendicular distances, and
3. To indicate group sizes (cluster sizes), separated by two spaces.

Each line should end with a semicolon. When there are no observations for a particular transect, there should only be a semicolon on that line. The Dist.dat data file for our warthog example, with 20 transects and 77 observations, will look as follows:

```
sample/effort=72.78;
102 2 42 2 58 1 29 2 98 1 5 1
29 3 21 1;
sample/effort=58.30;
47 4 20 4 49 1 6 2 42 1 39 1 85
3 0 2 32 1 0 1 243 2;
sample/effort=72.00;
11 3 0 2 8 2 50 1 57 1 0 1 7 3;
sample/effort=68.55;
2 3 33 1;
sample/effort=76.02;
;
sample/effort=74.65;
;
sample/effort=74.48;
;
sample/effort=74.53;
0 2 0 1 4 7;
sample/effort=63.62;
29 2 99 8;
sample/effort=62.23;
;
sample/effort=68.50;
;
sample/effort=62.36;
0 3 6 3 142 1 19 1;
sample/effort=65.12;
24 1 137 4 7 2;
sample/effort=73.93;
139 1 44 1 81 4 3 2 262 5 95 2
95 2 61 1 59 3 139 4 15 2;
sample/effort=74.53;
91 4 0 2 31 5 3 2 50 1 16 1;
sample/effort=75.37;
79 4 11 2 0 2;
sample/effort=75.21;
22 3 87 1;
sample/effort=75.93;
36 2 5 2;
sample/effort=24.00;
46 6 124 2 76 9 41 1 125 3 24 2
35 6 23 1 86 3 26 3;
sample/effort=51.75;
61 1 51 2 114 1;
end;
```

Note: the first line shows the length of the first transects combined (72.78 km), and the second line shows the observations in perpendicular distance (first observation is 102 m), followed by the group size (2 warthog (seen at 102 m)).

The interactive input for program DISTANCE for the first run should look as follows:

ENTER	EXPLANATION
Options;	Changes the prompt to OPTIONS>
Type=line;	Type of data from line-transect survey
Dist/units=metres;	Perpendicular distances are in metres
Area/units=sq. kilometres;	Area units are in square kilometres
Length/units=kilometres;	Transect lengths are in kilometres
Dist=perp;	Analysis uses perpendicular distances
Object=cluster;	Analysis uses groups of animals
Selection=all;	Fits all combinations of adjustment terms
Title=Warthog Density;	Title of run is "Warthog Density"
End;	Terminates the options section

Data; **Stratum /label='Lupande' /Area=1660;** **Infile=Dist.dat;**	Changes the prompt to DATA> Defines stratum and size Loads the data file
Estimate; **Est /key=hazard /adj=cosine;** **Est /key=hazard /adj=hermite;** **Est /key=hazard /adj=poly;** **Est /key=nexpon /adj=cosine;** **Est /key=nexpon /adj=hermite;**	Changes the prompt to ESTIMATE> Defines estimators and adjustment terms used for the analysis
End;	Starts analysis

The results, including two outliers, are summarised in Table 4.2. Program DISTANCE selected the Hazard estimator with Cosine adjustment terms as the best model fitting the perpendicular distance data for warthog.

Number of warthog in the survey area: N = 5,327 (SE = 42.84%)

95% Confidence Limits: %CL = 2,351 – 12,071

(Note that the 95% CL are asymmetrical, hence not based on a normal distribution of the data)

Table 4.2: Summary results of Lupande warthog data, analysed with program DISTANCE, including two outliers (>200 m). The Hazard estimator with Cosine adjustment terms was selected by DISTANCE, but the Negative Exponential estimator with Cosine adjustment terms gave the best estimate.

Estimator/Adjustment	Estimate	%SE	95% Confidence Limits
Hazard/Polynomial	3,524	40.73	1,612 – 7,702
Hazard/Cosine	**5,327**	**42.84**	**2,351 – 12,071**
Hazard/Hermite	3,480	39.81	1,618 – 7,486
N.Exponential/Cosine	2,304	33.45	1,192 – 4,453
N.Exponential/Polynomial	2,469	32.11	1,305 – 4,671
N.Exponential/Hermite	2,067	30.99	1,113 – 3,839
Uniform/Polynomial	1,150	30.38	624 – 2,118
Uniform/Cosine	1,606	33.27	833 – 3,096
Uniform/Hermite	987	-	3 – 383,150
Half Normal/Polynomial	1,275	30.24	694 – 2,343
Half Normal/Cosine	1,807	31.22	969 – 3,372
Half Normal/Hermite	1,302	30.04	711 – 2,386

The program selects the model that minimises the AIC, which does not necessarily mean that this model is the best choice, although many users may indiscriminately accept the estimator selected by DISTANCE. To avoid this, the output in the Dist.out file should be inspected carefully. If key parameters (*estimated coefficients, such as* a_l *and* a_2*; see Box 4.1*) are at lower or upper bound, and the sample size is large, the user may have to truncate the data. With small samples, truncation is not advisable, and the user may have to change the intervals used for constructing the frequency diagram with the GOF command.

The first step is to inspect the Effective Strip Width (ESW). As a rule of thumb, the ESW should not deviate too much from the actual mean perpendicular distance. When the ESW is low, the estimate will be high, and vice versa. The next step is to inspect the Estimated Cluster Size E(S). The E(S) should not deviate too much from the average cluster size, or the mean group size observed in the field. The last step is to inspect the AIC and the Chi-p of the model selected by the program, and compare this with the other models.

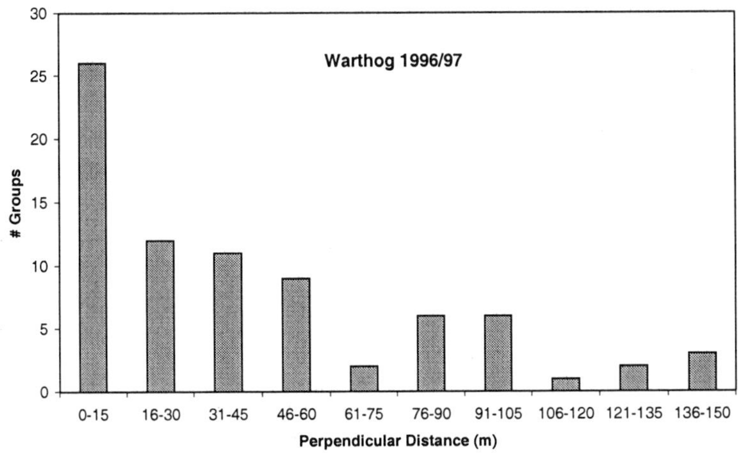

Figure 4.15: The perpendicular distance data for warthog, omitting two outliers, obtained through a series of 6 line-transect surveys in the Lupande GMA in 1996 and 1997. The data probably come from an underlying negative exponential detection function.

Based on the lowest AIC value, the program selected the Hazard estimator with Cosine adjustment terms (AIC=754.87), resulting in 5,327 warthog (Table 4.2). However, one out of 3 key parameters is at lower or upper bound, giving a biased estimate. This results in an Effective Strip Width (ESW) of 19.49 m, which compares with a mean perpendicular distance of 49.84 m, and an estimate that is too high. Also, the estimated function deviates too much from the frequency diagram of perpendicular distance data (P<0.026). The Negative Exponential estimator with Cosine adjustment terms, however, provides a better fit (Table 4.2), with an AIC that is only slightly higher (AIC=756.92). This estimator has no key parameters at lower or upper bound, and the estimated function provides a good fit, resulting in an ESW of 46.70 m, which is close to the mean perpendicular distance (49.84 m). Size-bias tests are non-significant. This estimator results in 2,304 warthog (Table 4.2).

When we omit the two outliers of 243 m and 262 m, using 10 classes of 15 m each (Figure 4.15), program DISTANCE selects the Hazard estimator with Cosine adjustment terms as the best model (Table 4.3).

Number of warthog in the survey area: N = 7,268 (SE = 43.69%)

95% Confidence Limits: %CL = 3,161 – 16,705

(Note that the 95% CL are asymmetrical, hence not based on a normal distribution of the data)

The Negative Exponential estimator with Cosine adjustment terms gives a similar result (2,215 warthog; *see Table 4.3*) as the preceding analysis with the two outliers (2,304 warthog). However, primarily based on the lowest AIC value, the program selects the Hazard estimator with Cosine adjustment terms, which gives an estimate of 7,268 warthog. This estimate is almost three times higher (Table 4.3). Here, the problem is mainly with small sample sizes, but also with limitations in the program for pooling the perpendicular distance data. As a consequence, 2 out of 3 key parameters could not be estimated, because they were at upper or lower bound, giving a biased estimate. The Negative Exponential estimator with Cosine adjustment terms has no key parameters at upper or lower bound. The ESW for the Hazard estimator with Cosine adjustment terms is 13.41 m. This is much lower than the mean perpendicular distance of 44.44 m. The ESW for the Negative Exponential estimator with Cosine adjustment terms is 47.85 m, which is a closer estimate. In this example, the E(S) does not show much difference from one model to the next, while size-bias tests were non-significant. Also, when we changed the intervals of the frequency diagrams of perpendicular distance data, using the GOF command, the results remained the same. The estimated function for the frequency diagram of perpendicular distance data gives a much better fit for the Negative Exponential estimator than for the Hazard estimator, with an AIC value that is only slightly higher.

Because the Hazard estimator with Cosine adjustment terms has 2 key parameters at lower or upper bound, resulting in an estimated strip width that is unrealistically low, and therefore a biased estimate, we select the Negative Exponential estimator with Cosine adjustment terms as the best model fitting the perpendicular distance data for warthog (Table 4.3).

Table 4.3: Summary results of Lupande warthog data, analysed with program DISTANCE, omitting two outliers (>200 m). The Hazard estimator with Cosine adjustment terms was selected by the program, but the Negative Exponential estimator with Cosine adjustment terms gave the best estimate.

Estimator/Adjustment	Estimate	%SE	95% Confidence Limits
Hazard/Polynomial	3,594	39.75	1,673 – 7,723
Hazard/Cosine	**7,268**	**43.69**	**3,162 – 16,705**
Hazard/Hermite	8,323	43.89	3,609 – 19,198
N.Exponential/Cosine	2,215	32.17	1,171 – 4,190
N.Exponential/Polynomial	3,008	34.64	1,538 – 5,883
N.Exponential/Hermite	-	-	-
Uniform/Polynomial	1,915	31.71	1,019 – 3,597
Uniform/Cosine	1,838	31.62	980 – 3,447
Uniform/Hermite	1,208	29.48	666 – 2,192
Half Normal/Polynomial	1,644	30.38	894 – 3,026
Half Normal/Cosine	1,795	31.48	959 – 3,360
Half Normal/Hermite	1,762	31.82	936 – 3,318

4.6.2 Analysing the Lupande Elephant Data with Program DISTANCE

The following example uses the elephant data collected in the Lupande GMA in 1997. There were 26 observations of elephant groups. Although the sample is too small for analysis with program DISTANCE, this exercise merely illustrates potential pitfalls when using small samples. Twenty block transects, each consisting of three segments, were counted three times in October. For the purpose of this example, each block transect was considered a single transect, while the 3 repeated counts for each block transect were pooled to improve precision. The perpendicular distances ranged from 0 m to 506 m, with 3 observations between 200 m and 250 m, and one at 506 m (*outliers*). The most convenient number of classes is 6 of 40 m each, omitting the extreme outlier of 506 m (Figure 4.16).

The frequency diagram shows us that the perpendicular distance data probably come from an underlying negative exponential detection function.

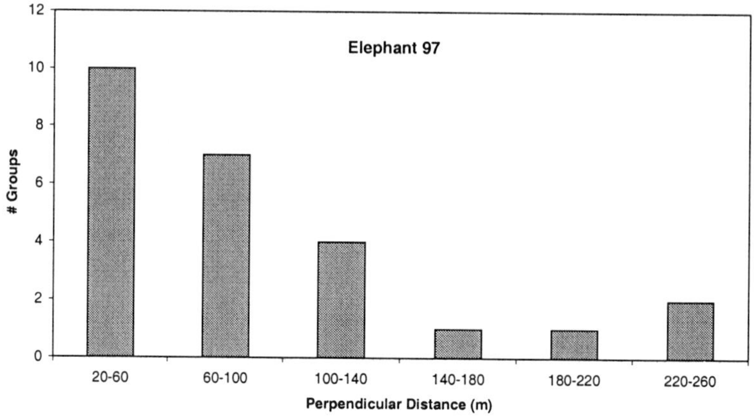

Figure 4.16: The perpendicular distance data for elephants, collected in the Lupande GMA in 1997, probably come from an underlying negative exponential detection function.

The Dist.dat data file for our elephant example, with 20 transects and 26 observations, will look as follows:

```
sample/effort=34.10;
7 4;
sample/effort=30.66;
;
sample/effort=35.23;
88 1 20 6 28 10;
sample/effort=34.85;
;
sample/effort=36.59;
78 5;
sample/effort=39.12;
;
sample/effort=38.80;
;
sample/effort=38.34;
;
sample/effort=40.79;
116 22 114 16;
sample/effort=37.57;
;
sample/effort=37.04;
;
```

```
sample/effort=37.48;
;
sample/effort=35.15;
;
sample/effort=40.27;
43 11 51 3;
sample/effort=38.11;
;
sample/effort=40.11;
25 2;
sample/effort=38.65;
0 3 3 1;
sample/effort=36.97;
;
sample/effort=19.50;
50 1 0 6 50 1 216 1 144 2 78 1
246 8 22 8 10 1 60 4 228 3 0 5
95 1 506 3;
sample/effort=38.29;
;
end;
```

The results are summarised in Table 4.4. Program DISTANCE selected the Negative Exponential estimator with Cosine adjustment terms as the best model.

Number of elephants in the survey area: N = 2,207 (SE = 81.40%)

95% Confidence Limits: %CL = 511 – 9,534

Table 4.4: Summary results of Lupande elephant data, analysed with program DISTANCE, including four outliers (>200 m). The Negative Exponential estimator with Cosine adjustment terms provided the best fit for the detection function.

Estimator/Adjustment	Estimate	%SE	95% Confidence Limits
Hazard/Polynomial	2,265	76.45	590 – 8,703
Hazard/Cosine	2,265	76.45	590 – 8.703
Hazard/Hermite	2,265	76.45	590 – 8,703
N.Exponential/Cosine	2,207	81.40	511 – 9,534
N.Exponential/Polynomial	2,150	81.90	494 – 9,354
N.Exponential/Hermite	1,877	80.11	440 – 8,006
Uniform/Polynomial	1,022	74.02	275 – 3,798
Uniform/Cosine	1,416	73.61	383 – 5,231
Uniform/Hermite	743	68.52	216 – 2,556
Half Normal/Polynomial	1,013	69.89	289 – 3,554
Half Normal/Cosine	1,516	71.48	423 – 5,434
Half Normal/Hermite	1,093	69.63	313 – 3,820

When we omit the four outliers (Figure 4.17), program DISTANCE selects the Hazard estimator with Cosine adjustment terms as the best model (Table 4.5).

Number of elephants in the survey area: N = 14,646 (SE = 84.16%)

95% Confidence Limits: %CL = 3,402 – 63,058

Considering the fact that annual aerial sample counts, carried out in the Lupande GMA between 1993 and 1999, returned fewer than 1,000 elephants, the above estimate seems much too high.

The Negative Exponential estimator with Cosine adjustment terms gives a similar result as the analysis with the four outliers. The program's choice of Hazard plus Cosine was based on a slightly lower AIC value (213.56), compared to that for the Negative Exponential model (213.83).

However, the model selected by the program has 2 out of 3 key parameters at upper or lower bound, and this explains the biased estimate.

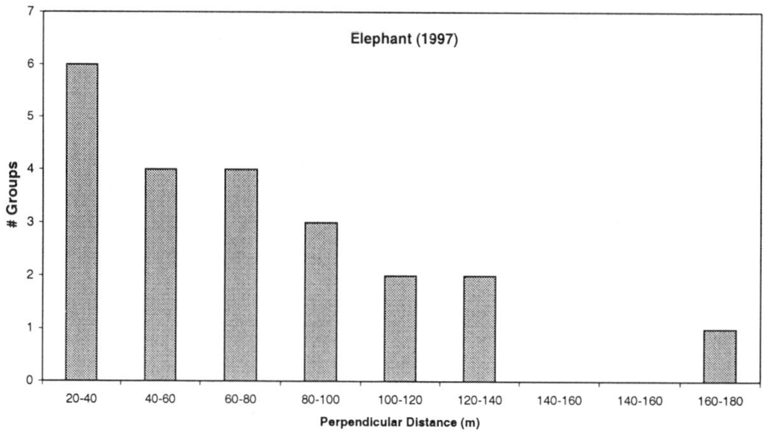

Figure 4.17: The perpendicular distance data collected for elephants in the Lupande GMA in 1997, omitting 4 outliers, probably come from an underlying negative exponential detection function.

The ESW for the Hazard estimator with Cosine adjustment terms is only 12.89 m, which compares with an actual mean strip width of 49.31 m. The ESW for the Negative Exponential estimator with Cosine adjustment terms is 61.65 m, which is closer. The E(S) for the Hazard model with Cosine adjustment terms is 7.50, which compares with an average cluster size of 5.18. The E(S) for the Negative Exponential model with Cosine adjustment terms is 5.77, which is closer. The estimated function for the frequency diagram of perpendicular distance data has a much better fit for the Negative Exponential estimator than for the Hazard estimator. Size-bias tests are non-significant. Different intervals for the frequency diagrams of perpendicular distance data give similar results. Also, the Negative Exponential estimator has no key parameters at upper or lower bound. Based on these criteria, we decide to use the Negative Exponential estimator with Cosine adjustment terms as the best model fitting the perpendicular distance data for elephants.

In summary, with small samples of < 80 observations, it is advised to include the least extreme outliers in the analysis. When outliers are omitted, careful inspection of the output is required. Even when using default settings, especially with small sample sizes, one cannot indiscriminately accept the estimator selected by DISTANCE.

We have demonstrated that low wildlife densities result in too few sightings for analysis with program DISTANCE. Most protected areas on the African continent face similar problems. With small sample sizes, a simple alternative short-cut method is to use twice the mean perpendicular distance as the mean transect width. Although this method only requires the use of a calculator and some common sense, it is not always valid. For large samples the results are never as accurate as those obtained with conventional line-transect mathematics. We will continue with a brief discussion of the theoretical concept, and the conditions under which it is appropriate.

Table 4.5: Summary results of Lupande elephant data, analysed with program DISTANCE, omitting four outliers (>200 m). The Hazard estimator with Cosine adjustment terms was selected by the program, but the Negative Exponential estimator with Cosine adjustment terms gave a better fit.

Estimator/Adjustment	**Estimate**	**%SE**	**95% Confidence Limits**
Hazard/Polynomial	3,533	70.12	919 – 13,585
Hazard/Cosine	**14,646**	**84.16**	**3,402 – 63,058**
Hazard/Hermite	17,797	84.34	4,123 – 76,811
N.Exponential/Cosine	2,349	74.49	609 – 9,060
N.Exponential/Polynomial	2,349	74.49	609 – 9,060
N.Exponential/Hermite	2,349	74.49	609 – 9,060
Uniform/Polynomial	1,246	60.66	410 – 3,792
Uniform/Cosine	1,547	61.70	500 – 4,781
Uniform/Hermite	1,497	61.52	486 – 4,615
Half Normal/Polynomial	1,528	62.19	491 – 4,755
Half Normal/Cosine	1,528	62.19	491 – 4,755
Half Normal/Hermite	1,528	62.19	491 – 4,755

4.7 The Mean Perpendicular Distance Method

The Mean Perpendicular Distance Method (MPD) uses line-transect methodology, but not line-transect mathematics. By using twice the mean perpendicular distance, the population estimate and precision can be calculated without a computer. We have mentioned that fixed-width transects are a special case of line transects (*see section 4.1.1*). For fixed-width transects, the estimator of the density (D) of (n) observations of animals or groups of animals over a transect line with length (L) and ½ width (w) is:

$$D = \frac{n}{2w \times L} \quad (1)$$

Thus, if we count 100 groups of animals (n) on 10 fixed-width transects with a total line length of 100 km (L), and a width of 0.1 km on either side (½w), the density (D) is:

$D = 100/(0.2\text{km} \times 100\text{km}) = 5$ groups of animals per km².

If we count 100 groups of animals (n) on 10 line transects with a total line length of 100 km (L), and a mean perpendicular distance of 0.1 km on either side ($\frac{1}{2}p_m$), the result is the same, with the only difference that ½w is replaced by $\frac{1}{2}p_m$. Thus, using the mean perpendicular distance in line-transect methodology is a special case of fixed-width transects, essentially transforming a variable visibility profile into a fixed width strip. The field methodology is the same as for line transects analysed with DISTANCE.

The last thing we would like to do, however, is to promote the use of this method under all circumstances. It is strongly advised to apply line-transect mathematics, by using the appropriate software, whenever possible. However, with small sample sizes (few sightings), and where computer, software and expertise are not available, the MPD method provides an alternative for analysing line-transect data, as long as its weaknesses are well understood. Following the next section on analysis with the MPD method, the two previous examples analysed by DISTANCE will be analysed with the MPD method.

4.7.1 Data Analysis using the Mean Perpendicular Distance Method

Analysis by the mean perpendicular distance method (MPD) follows a series of logical steps.

1. Convert the sighting bearings to sighting angles relative to the line of travel and calculate the perpendicular distance for each observation,

2. Calculate the total line length covered,

3. Calculate the total number of individuals observed for the species in question,

4. Calculate the mean perpendicular distance as explained below,

5. Determine the population estimate and the level of precision as explained below.

Consider a sample of *m* transects with *n* sightings of groups of animals. We can work up the results as follows:

1. The mean perpendicular distance (p_m) is simply calculated by dividing the enumerated distances i (i is the first sighting) to n (Σp_i in km) by the number of groups counted (n):

$$p_m = \frac{\Sigma p_i}{n} \quad \text{(km)}$$

2. This is half the effective strip width, which has to be multiplied by factor 2 to obtain the full strip width:

$$\frac{\Sigma p_i}{n} \times 2 \quad \text{(km)}$$

3. Multiplying the effective strip width with the total length of i (is first transect) to m transects combined (ΣL_i in km) gives the total area sampled (A_s in km²):

$$A_s = \frac{\Sigma p_i}{n} \times 2 \times \Sigma L_i$$

4. The density of groups of the target species (D_g) can now be estimated by dividing the total number of groups observed (n) in m transects by the total area sampled (A_s):

$$D_g = \frac{n}{\dfrac{\Sigma p_i}{n} \times 2 \times \Sigma L_i} = \frac{n}{A_s}$$

5. Multiplying the density of groups (D_g) by the area (A_t) gives the total number of groups in the survey area (D_t):

$$D_t = D_g \times A_t$$

6. The size of the population (N) can be estimated by multiplying the total number of groups (D_t) by the mean group size (G_m):

$N = D_t \times G_m$ (where $G_m = \Sigma b/n$; Σb is the total number of animals observed)

The formula for calculating the population estimate with twice the mean perpendicular distance is shown in Box 4.2.

Box 4.2: Formula for population estimate, using twice the mean perpendicular distance.

Combining 1 to 6 gives:

$$N = \frac{A_t \times \Sigma b}{\dfrac{\Sigma p_i}{n} \times 2 \times \Sigma L_i}$$

where: N is the population estimate,
Σp_i is the enumerated perpendicular distances of i to n sightings,
Σb is the total number of individual animals counted,
ΣL_i is the total length of i to *m* transects (km),
n is the number of sightings of groups of the target species,
A_t is the area of the survey zone (km²).

The variance can be estimated directly by estimating the population size for each individual transect. Direct estimation of variance is only possible when each transect has a sufficient number of observations. When the sample contains transects with few observations or none at all, variance should either be estimated indirectly, or the data of several transects running parallel to each other should be pooled. When each transect contains a sufficient number of observations, direct estimation of variance is a straightforward procedure:

An overall, weighted estimate of population size (N_w) is obtained by (Burnham *et al.,* 1980):

$$N_w = \frac{\sum_{i=1}^{m} L_i N_i}{L}$$

where:

L_i = Length of ith transect (i to m transects in sample),
N_i = Population estimate for ith transect,
L = Length of all transects combined,
m = Number of transects covered during the survey.

Estimation of variance is as follows:

$$Var(N_w) = \frac{\sum_{i=1}^{m} L_i (N_i - N_w)^2}{L\,(m-1)}$$

where:

Var = Variance,
N_w = Weighted estimate of population size.

The Standard Error is $\sqrt{Var}$

The 95% Confidence Limits can be approximated by t x SE, where t = 1.96 when m is large.

With low densities, however, there will always be a number of transects without observations. Then, data will have to be pooled. With a sample size of 20 transects, one can pool the data in, for instance, 4 groups of 5 transects that run parallel to each other. The same formulae can be used (m = 4). The procedure is as follows:

a. Pool the data of several transects running parallel to each other to enable computation of at least 4 population estimates (N_i),

b. Estimate each population size of the pooled data per group of transects (N_i) with the MPD method,

c. Compute the weighted estimate of population size (N_w),

d. Estimate the variance with the formula provided above.

The Standard Error is $\sqrt{Var}$

The 95% Confidence Limits can be approximated by t x SE, where t = 1.96 when m is large.

4.7.2 Analysing the Lupande Warthog Data with the MPD Method

The warthog data from the Lupande GMA (number of sightings is 77) will be used to illustrate the MPD method, by following steps 1 to 6 and a to d above. The warthog data come from an underlying negative exponential detection function (Figure 4.14).

1. $\frac{\Sigma p_i}{n} = \frac{3{,}838}{77} = 49.84$ m/1,000 = 0.04984 km (mean perpendicular distance)

2. 0.04984 x 2 = 0.09968 km (twice the mean perpendicular distance)

3. A_s = 0.09968 x 1,344 (total transect length in km) = 133.97 km² (total area sampled)

4. $D_g = \frac{77}{133.97} = 0.575$ groups of warthog/km² (group density)

5. D_t = 0.575 x 1,660 km² (size of survey area (A_t)) = 954 groups in the survey area

6. $N = 954 \times \frac{185}{77}$ (mean group size (Gm)) = 2,292 warthog (population estimate)

The weighted estimate of population size (N_w) and estimation of variance are computed as follows:

a. Pool the data of transects 1 – 5, 6 – 10, 11 – 15 and 16 – 20,

b. Compute 4 population estimates N_i, by executing steps 1 to 6 (results: $N_1 = 2{,}023$; $N_2 = 1{,}799$; $N_3 = 2{,}200$; $N_4 = 3{,}002$),

c. Compute the weighted estimate of population size (result: $N_w = 2{,}230$; $m = 4$),

d. Compute $Var(N_w)$, (result: $Var\ N_w = 645{,}590$),

Standard Error is $SE(N_w) = 803$

The final estimate is 2,230 ± 36.01% (Y ± SE), which compares well with 2,304 ± 33.45% (Y ± SE), obtained by using the Negative Exponential estimator with Cosine adjustment terms. The MPD result is only 3.2% lower than the result from DISTANCE.

4.7.3 Analysing the Lupande Elephant Data with the MPD Method

The same analysis will be performed with the elephant data from the Lupande GMA (number of sightings is 26). These data also come from an underlying negative exponential detection function (Figure 4.16).

However, due to a highly uneven distribution of elephants in the Lupande GMA, only a few transects contain observations, while many transects have zero elephants. Estimation of variance is done indirectly, by estimating separately the sampling variances of the number of groups observed (n) and the reciprocal of the strip width ($^1/_w$) as follows:

$$Var(N_w) = L \frac{\sum_{i=1}^{m} L_i \left(\frac{n_i}{L_i} - \frac{n}{L} \right)^2}{m-1} \quad \text{(Burnham \textit{et.al.}, 1980)}$$

$SE(n) = \sqrt{Var\ (n)}$

$$CV(n) = \frac{SE(n)}{n}$$

$$Var(^1/w) = \frac{\sum_{i=1}^{m} \left(^1/w_i - ^1/w_m \right)^2}{m - 1}$$

$$SE(^1/w) = \sqrt{Var(^1/w)}$$

$$CV(^1/w) = \frac{SE(^1/w)}{w_m}$$

$$Var(N_w) = N_w^2 \{(cv(n))^2 + (cv(^1/w))^2\}$$

where:

L = Length of all transects combined,
L_i = Length of ith transect,
n_i = Number of groups observed in ith transect,
n = Number of groups observed in all transects combined,
m = Number of pooled groups of transects,
w_i = Mean perpendicular distance for ith group of transects,
w_m = Mean perpendicular distance for all transects combined,
CV = Coefficient of variation,
N_w = Weighted estimate of population size.

1. $\frac{\Sigma p_i}{n} = \frac{2{,}278}{26} = 87.61$ m/1,000 = 0.08761 km (mean perpendicular distance)

2. 0.08761 x 2 = 0.17523 km (twice the mean perpendicular distance)

3. A_s = 0.17523 x 727.62 (total transect length in km) = 127.51 km² (total area sampled)

4. $D_g = \frac{26}{127.51} = 0.204$ groups of elephants/km² (group density)

5. $D_t = 0.204 \times 1{,}660$ km² $(A_t) = 339$ groups in the survey area

6. $N = 339 \times \frac{129}{26} = 1{,}682$ elephants

The weighted estimate of population size (N_w) and estimation of variance are computed as follows:

a. Pool the data of transects 1 – 5, 6 – 10, 11 - 15, and 16 – 20,

b. Compute 4 population estimates N_i, by executing steps 1 to 6 (results: $N_1 = 2{,}855$; $N_2 = 1{,}409$; $N_3 = 1{,}315$ and $N_4 = 2{,}393$),

c. Compute the weighted estimate of population size (N_w), (result: $N_w = 1{,}960$),

d. Compute Var(N_w), (result: Var(N_w = 20,524,747),

e. Standard Error is SE(N_w) = 1,433

The final estimate is 1,960 ± 73.11% (Y ± SE), which compares well with 2,207 ± 81.40% (Y ± SE), obtained by using the Negative Exponential estimator with Cosine adjustment terms. The MPD result is only 11.2% lower than the result from DISTANCE.

The MPD method gives reliable results when the data come from an underlying negative exponential detection function, which has a simple estimator of f(o). The MPD method may also give reliable results for data coming from an underlying reverse logistic or uniform detection function, but when the data come from an underlying half normal detection function (few cases), MPD may give overestimates.

An example from the Nazinga Game Ranch in Burkina Faso will further illustrate the validity of the MPD method. Between 1987 and 1989, elephant abundance was established by an aerial total count (366 elephants), a series of line-transect surveys and a road survey (Jachmann, 1991a). The line-transect counts used the Fourier Series and the Modified Hayne estimators to model the detection function, resulting in 487 and 306 elephants respectively. In spite of the fact that road surveys usually do not cover a representative part of the study area, analysis of the road survey with the MPD method produced an estimate of 293 elephants, i.e. only 20% lower than the aerial total count. The Fourier Series overestimated by 33% and the Modified Hayne estimator underestimated by 16% (Jachmann, 1991a). Again, this demonstrates the usefulness of the MPD method. When sample sizes are large enough and you have the necessary mathematical skills, then you should use DISTANCE or LOPES. But if your sample size is too small, or you lack a computer or the necessary skills, then the MPD method is a simple alternative (*also see next section*).

4.8 Vehicle Counts

As with most direct counts, vehicle counts can be divided into total counts and sample counts. Total counts by vehicle should not be considered a valid option because of the reasons outlined in a previous chapter.

Sample counts by vehicle can be divided into off-road counts and road counts. Off-road counts are used in wide open country with good visibility, but only in conservation areas where animals are used to vehicles. They employ the same principles as sample counts on foot. Thus, when using line-transect methodology, statistical design procedures and analysis are exactly the same as those used for foot counts. The only difference is the use of a mechanical aid to cover the transects.

4.8.1 Road Counts

Road counts, however, are restricted to the existing road-system and therefore have different statistical design requirements than regular sample counts on foot or by vehicle. But the analysis is more or less the same as for conventional line-transect methodology. Although roads in conservation areas usually do not follow straight lines, a road survey can be considered as a special case of a line-transect survey with transect lines having random starting points, length and direction. At each point of the curved road, a tangential baseline can be imagined, from which sighting angles can be measured accurately.

However, there are certain problems with road counts, mainly concerned with the statistical design. Road counts are open to bias because the road-system in most conservation areas is unlikely to be representative (Norton-Griffiths, 1978). The road-system has usually been constructed to cover the most scenic parts with high densities of wildlife, frequently running alongside rivers and large open dambos with relatively rich soils. To facilitate construction and to limit costs, roads are usually built along rather than across contours (Norton-Griffiths, 1978). In addition, roads tend to be utilised by certain species such as elephant and impala, that graze the annual grasses along the verges. Pronounced seasonal movements of wildlife, such as the dry-season concentrations of many species near water sources, will change the bias in the course of the year, resulting in overestimates in the dry season and underestimates during the wet. Failure to take these factors into consideration when designing a road count will lead to biased results.

If the road-system does not cover all parts of a particular conservation area, most sources of bias can be avoided by stratification, for instance by vegetation type. The first step is to estimate the densities of the target species for each vegetation type. The second step is to multiply these densities by the surface area of each vegetation type in the entire study area. The third and last step is to add these together to give estimates of the total numbers in the whole study area.

However, when the road-system primarily runs through a single vegetation type, such as alongside a major river, road counts should be conducted at the peak of the dry season only, covering a single vegetation stratum. Here we assume that the majority of animals of the target species are found within that particular stratum. When the road-system runs through a single vegetation type that does not include a major river where animals concentrate during the peak dry season, a road count will give biased results.

As a general rule, in areas without a proper road-system, road counts should be avoided. If budgetary constraints make them the only option, great care must be taken to eliminate the sources of bias inherent in the method.

Where an extensive road-system is more or less representative of the area, road counts may produce excellent results at minimum expense. A good example is the multi-species road count conducted on the Nazinga Game Ranch in Burkina Faso in 1988, in conjunction with a conventional line-transect count on foot (Jachmann, 1988a).

The Nazinga Game Ranch covers approximately 940 km² in south-central Burkina Faso. The landscape is slightly undulating, traversed by three major rivers, the Sissili, the Dawevele and the Nazinga. The vegetation is a tall-grass shrub savanna with *Terminalia*, *Combretum* and *Acacia* species as the dominant woody plants (Jachmann and Croes, 1991). Annual rainfall averages 880 mm, falling mainly between June and September.

The line-transect foot count was conducted during the first week of April 1988. Nine large ungulate species were counted, e.g. bushbuck (*Tragelaphus scriptus*), elephant (*Loxodonta africana*), Grimm's duiker (*Sylvicapra grimmia*), hartebeest (*Alcelaphus buselaphus major*), oribi (*Ourebia ourebi*), reedbuck (*Redunca redunca*), roan antelope (*Hippotragus equinus*), warthog (*Phacochoerus aethiopicus*) and waterbuck (*Kobus defassa*). Sixty-five transects, totalling 562.1 km, took 5 days to complete with 13 teams of 3 observers. The extensive infrastructure on the ranch enabled a simple systematic lay-out of transects, each beginning and ending on a road. Density estimates were made using the Fourier Series estimator when the number of observations of groups of animals exceeded 25, and the Modified Hayne estimator for those species with more than 6 but less than 25 observations. In the previous chapter we showed that most perpendicular distance data of large African ungulates come from an underlying negative exponential or reverse logistic detection function g(x), for which the Fourier Series estimator may not be the appropriate choice to model the detection function. However, in the late 1980s, when we carried out the surveys on the Nazinga Game Ranch, the Fourier Series estimator was considered the most robust and best option available (Burnham *et al.*, 1980).

The road survey was conducted during the last week of March 1988. A total of 226.3 km was driven, covering most of the road-system in the ranch, at a mean speed of 11.8 ± 2.6 km per hour. The perimeter roads were excluded. The survey was done with a pick-up truck, with three observers standing in the back. Two observers spotted and counted the animals, each on a different side of the road, and one observer measured the distance and sighting angle using a range-finder and protractor. The road survey data were analysed in two different ways, first by using twice the mean perpendicular distance as the mean transect width, and second by using a modified version of the method of Anderson and Pospahala (1970), which will be discussed below.

The data from the foot counts were analysed with program TRANSECT. For the foot counts, the precision of the population estimates was provided by program TRANSECT, whereas the precision of the road count data was obtained by using replicate lines (Burnham *et al.*, 1980). When using replicate lines to determine the variation in the results of a road count, certain stretches of road, representative of the whole study area, are counted repeatedly.

The results of the line-transect foot count are very similar to those obtained from the road count (Table 4.6). What can we learn from this? Firstly, in conservation areas with an extensive road-system, comprising a representative sample of the area, road counts are a valid and cost-efficient option. The fact that it is also cost-efficient will be discussed in the last part of this chapter. Secondly, the use of twice the mean perpendicular distance as the mean transect width is a simple and valid alternative to computers and complicated software.

The third way to analyse the road count data as well as any other line-transect data uses a modified version of the method of Anderson and Pospahala (1970). Using this method, at least 25 sightings per species are required for constructing a frequency diagram of perpendicular distances (Jachmann, 1988a). Assuming that no animals on or near the transect line are missed, the frequency diagram reflects the detection curve g(x). Sightings having extreme perpendicular distances are called outliers and can be eliminated from the analysis. Transect width is determined by the maximum perpendicular distance of the frequency diagram. Hence, using this particular method, it is far more important to eliminate outliers than in any of the previous methods of analysis discussed above.

As an example, we will analyse the oribi data using this method. Analysis follows a simple graphical procedure (Figure 4.18). The sighting in the range between 180 and 200 m is eliminated as an outlier, giving a maximum effective transect width of 140 m. The area of the rectangle below the detection curve represents the number of oribi sighted (Figure 4.18). Assuming that within the transect densities of oribi were uniform, then the area above the detection curve represents the number of oribi missed during the survey. Thus, the area of the rectangle represents the total number of oribi within the transect. When there is no statistical program available to calculate the detection curve, the curve can be approximated by drawing a smooth line through the means of two consecutive data points in the frequency diagram (Figure 4.18). The next step is to plot the frequency diagram on graph paper and to count the squares (Figure 4.18). The number of animals observed on the transect can now be corrected for the animals missed during the survey by multiplying with the total area of the rectangle divided by the area below the detection curve.

In our example (Figure 4.18), the correction factor is roughly 120 squares (entire rectangle) divided by 36 squares (area below detection curve) is 3.3. Thus, to calculate the population estimate, the number of oribi observed in the transects should be multiplied by factor 3.3. The estimate was 1,869 ± 1,839 animals, although very imprecise, only 23% lower than the estimate from the mean perpendicular distance, and 16% higher than the estimate using the Fourier Series estimator (Table 4.6).

Table 4.6: Summary results of a line-transect foot count analysed with program TRANSECT and a road count conducted on the Nazinga Game Ranch in 1988 (Population Estimates with Standard Errors).

SPECIES	FOOT COUNT	ROAD COUNT
	Population Estimates	**Population Estimates**
	Program Transect	**Mean Perp. Dist.**
Bushbuck	942 ± 582	573 ± 817
Duiker	663 ± 357	608 ± 460
Elephant	306 ± 646	293 ± 222
Hartebeest	1,011 ± 1,150	835 ± 631
Oribi	1,611 ± 561	2,432 ± 1,207
Reedbuck	109 ± 167	100 ± 76
Roan	1,650 ± 862	1,743 ± 1,318
Warthog	4,736 ± 1,539	5,139 ± 3,813
Waterbuck	231 ± 378	234 ± 148

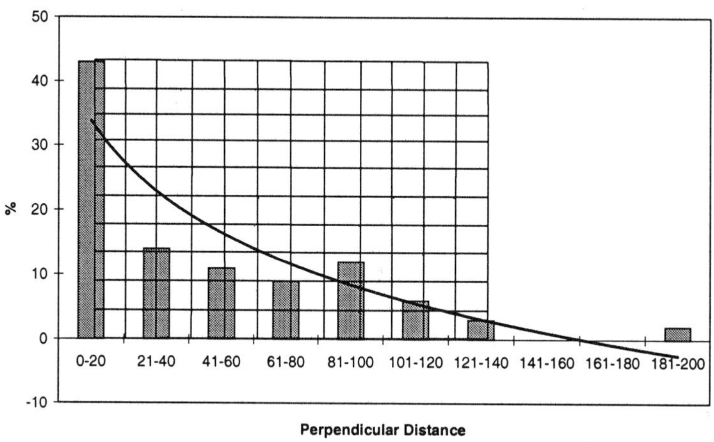

Figure 4.18: Perpendicular distance data for oribi, collected during a road survey on the Nazinga Game Ranch in 1988. The frequency diagram shows the detection function g(x) and a grid that can be used to estimate the fraction of oribi missed on the transect. Observations made at perpendicular distances from the transect line of more than 140 m (the width of the transect) are discarded as outliers.

4.9 Comparative Costs of Sample Counts on the Ground

The cost of a sample count on the ground is a function of the type of count, local wages and transport costs, the accessibility of the area and above all the objective of the count, and therefore the precision required for the population estimates. The level of precision is directly related to the spatial distribution and the abundance of the target species. Thus, in areas where the target species is distributed more or less evenly, the number of sightings determines the precision of the estimates. At this stage it may be useful to compare the costs of a line-transect survey conducted in Nazinga in 1988 and in the Lupande GMA in 1996 (Table 4.7).

First of all, the mean standard error for 5 common ungulate species counted during the single line-transect survey in Nazinga was 104.58%. The mean standard error of the results for the same species in the Lupande GMA was 43.94%. In Nazinga, the coverage was 562.1 km of transect, divided over 65 transects, in a study area of 940 km². In the Lupande GMA the coverage was 625.1 km of transect, divided over 20 transects sampled 6 times, in a study area of 1,660 km². Although there is not much difference in the total length of transect, the higher level of precision obtained in the Lupande GMA is a consequence of using twice the number of transects. Hence, it is much better to use a large number of short transects than a small number of long transects. In addition, two of the species showed a more clumped distribution at Nazinga.

In Nazinga, the cost of a single line-transect survey was US$4,500, or US$8.0 per km of transect, or US$4.8 per km². In the Lupande GMA, the cost of a series of 3 surveys was US$7,500, or US$12 per km of transect, or US$4.5 per km² (Jachmann, 1988a). The higher cost per km of transect in Lupande is mainly due to problems of access and a larger study area in combination with repeated coverage, whereas the extensive road-system in Nazinga facilitated logistics involved with the survey.

The cost of the single road survey conducted on the Nazinga Game Ranch was US$500, or US$ 2.2 per km of road, or US$0.5 per km². This does not include maintenance of roads, because this is done for other reasons, but it does include vehicle replacement. This is substantially lower than the costs of a line-transect survey, producing similar results with approximately the same levels of precision. Therefore, in conservation areas with an extensive road-system that is representative of the whole area, the preferred choice for monitoring wildlife abundance is the road count, using twice the mean perpendicular distance as the mean transect width.

Table 4.7: Comparative costs of line-transect surveys and road surveys.

	NAZINGA		LUPANDE
Survey Area	940 km²		1,660 km²
Number of Transects	65		120
Transect Length	562.1 km		625.1 km
Mean Standard Dev.	104.58%		43.94%
	LINE TRANSECT	**ROAD**	**LINE TRANSECT**
Total Costs	US$4,500	US$500	US$7,500
Costs per km Transect	US$8.0	US$2.2	US$12.0
Cost per km²	US$4.8	US$0.5	US$4.5

PART IIb

AERIAL COUNT TECHNIQUES

The potential of low-altitude flying to collect habitat information was first recognised in the early 1920s, when aerial photographs were taken in East Africa by the Royal Air Force. Then, the objective was to collect general information on habitats and wildlife. In the late 1940s, light aircraft were used for ecological information gathering, spotting wildlife and as assistance in law-enforcement operations. It was not until the mid and late 1950s that park wardens and wildlife researchers began to use light aircraft in attempts to obtain total counts of wildlife species in East Africa. Not long after these first attempts, biologists realised that it would be more efficient to use some form of sampling rather than counting all the animals in the target population (Watson *et al.*, 1968). Furthermore, it was becoming apparent that for management purposes information on population trends was probably more useful than total numbers. By the late 1960s, low-level aircraft surveys in eastern Africa were a common practise and had developed their own distinctive character, whereby information gathered from the air was frequently related to data collected on the ground. In the early 1970s, satellite remote sensing had partly replaced aerial techniques for monitoring habitat changes. Although aerial counting techniques have always been highly overrated with regard to accuracy, they continue to be one of the many tools in wildlife management.

As with ground counting techniques, aerial techniques can be divided into total counts or censuses and sample counts or surveys. The balance between intensive total counts on the one hand and low coverage sample counts on the other hand should be considered in relation to the objectives of the census or survey, the size of the study area and the potential uses of the data to be collected. As with any wildlife count, these objectives and uses must be made clear prior to conducting the count. Although the management objective should be the single most important factor determining the type and intensity of a particular count, practice dictates that resource limitations in terms of the available time, manpower and above all the size of the budget usually bring down the management objective to a second-level priority. However, if resources are not adequate to achieve the objective, in the end it is cheaper and therefore wiser not to carry out a count that may give misleading results.

Since the 1950s, a large number of scientific papers have appeared on the subject of aerial counting, and the last thing we would like to do is to reinvent the wing. Instead, after a general overview of the theory and field procedures, we will attempt to clarify and simplify some important issues such as biases in the different techniques.

Chapter 5

AERIAL TOTAL COUNTS

The technique and the field procedures relating to aerial total counts were first summarised by Norton-Griffiths (1975). Although the concept is simple, the design of an aerial census requires careful consideration to minimise error and bias. The main objective of an aerial census is to describe accurately the total number of a particular target species, and its spatial distribution over the study area. The census requires at least two observers, each counting on a different side of the aircraft, to scan the entire study area, as the aircraft flies along parallel flight lines that are between 500 m and 2 km apart. In the case of a hippo count the observers scan rivers and pools, while for puku, reedbuck and oribi they scan dambos.

The distance between flight lines depends upon several factors. The target species, the searching rate (which is the area scanned per time unit), the height above ground level of the aircraft, the density of the vegetation and the topography of the study area are the most important ones.

Study areas larger than 1,000 km^2 should be divided into discrete blocks that can be covered easily by a single aircraft in one flying day. These blocks should be defined by features such as roads, water-sheds or topography.

In principle, a total count implies that there is no sample error attached to the final estimate of numbers. Unfortunately this has led to the rather uncritical acceptance of census figures, because researchers and managers have the tendency to overlook the fact that other sources of error and bias may generate unreliable results when an aerial census is not properly designed and conducted (Norton-Griffiths, 1978). A particular source of bias, known as visibility bias (see below), renders certain animal species unsuitable for any type of aerial count. A species such as black rhinoceros (*Diceros bicornis*) is impossible to count from the air with any level of precision or accuracy. In 1967 some experiments were conducted with a small rhino population, of which the numbers on the ground were known accurately, occupying the Olduvai Gorge in Tanzania (Goddard, 1967). The Olduvai Gorge is a wide open area with few scattered trees and bushes. Even under the most ideal conditions only 50% of the population was detected by observers in an aircraft. Goddard's conclusion was that light aircraft were of limited value in providing estimates of black rhino populations. Even animals as large as elephants may be missed at distances of over 100 m away from an aircraft flying at 100 m above ground. This simply means that an effective total count should be done at such a high intensity, that it becomes too costly to deploy. However, aerial total counts may give useful results for conspicuous species occurring in open terrain, such as zebra, buffalo, lechwe, wildebeest and elephant.

5.1 Sources of Bias in Aerial Total Counts

When designing an aerial total count, the following sources of bias require careful consideration (Figure 5.1):

1. **Insufficient coverage of census area**: This is a bias that may result in undercounts by as much as 40% (depending upon the target species), and is solely related to the design of the census. When the parallel flight lines are set too far apart apart, many animals will be missed. This is mainly the result of bias, and it increases with increasing distance between flight lines. It is a consequence of the observer having to scan too large an area on his side of the aircraft. When conducting a total count of elephants in the Luangwa Valley, the density of the woodland savanna dictates a maximum strip width of 500 m on either side of the aircraft, giving an interval of 1 km between parallel flight lines. When counting elephants in open flat terrain, parallel flight lines may be set apart at intervals of up to approximately 2 km. Thus the cost of an aerial census depends on the target species, the density of the vegetation, the topography and the size of the census zone.

2. **Visual estimation of large herds**: This is also a bias that may result in undercounts by as much as 30% (Norton-Griffiths, 1978), and is related to the availability of proper equipment to conduct the census. Regardless of the type and density of the vegetation, large herds should always be photographed. Even when the view of part of the herd is obstructed by canopy cover, the figures obtained from the photographs should be compared with those obtained by visual estimation. However, visual estimation should always be used as an insurance against photographic failure.

3. **Double counting of animals**: With the invention of the GPS, this became a less serious bias resulting from herds counted twice by

mistake. This bias is solely related to the design of the census. Generally, it will be negligible when using photography and when groups of animals counted are plotted on a map.

4. **Observer bias**: This may be a serious negative bias. It is related to the quality of the observers, which reflects the preparations of the census in terms of choice of census staff. The quality of the observers depends on their eye-sight and their capacity to count animals for prolonged periods under difficult conditions such as heat and turbulence, and their experience in terms of hours of counting the target species in similar types of habitat. Ability to concentrate during a long flight will vary from person to person, depending upon the observer's capacity to sit in a hot confined space and concentrate for hours at a time. Concentration is also affected by the length of the flight, which is determined by the census plan. Observer bias can be minimised by applying the double-count method (*see section 5.4*).

5. **Visibility bias**: Depending upon the target species and the density of the vegetation, there are always some animals that are hidden from the observers by obstructions such as tree canopies. In addition, in total counting there is always the problem of the strip directly below the aircraft. This affects most fixed-wing aircraft, but not so much helicopters. This bias may be reduced by applying the double-count method (*see section 5.4*).

6. **Sighting Probability Bias**: At first sight this bias appears to be similar to visibility bias. However, it does not relate to the invisible population of a particular target species, but to the low probability of spotting small groups of animals. Depending upon the species, the lower sighting probability is most apparent with solitary animals and groups of < 5 individuals. This bias becomes more apparent with an increase in searching rate and height of the aircraft (Caughley, 1974). Therefore, when designing a census, operational factors such as searching rate and height above ground level should be kept within reasonable limits.

What did we learn from the discussion on potential errors and biases when designing an aerial census? Sighting probability is a function of aircraft speed and height, abundance and average group size of the target species, and also of vegetation density, topography, light conditions, colour patterns of the animals to be counted and several unknown factors.

Bias resulting from insufficient coverage of the census area (1), visual estimation of large herds (2) and double counting of herds (3) can easily be reduced or avoided with a proper census design. Observer bias (4) can be reduced by using experienced staff, while the double-count procedure (*see below*) will account for part of the other animals missed. Visibility bias (5), however, cannot be estimated and therefore cannot be corrected. However, use of the double-count method will correct for part of the animals missed due to visibility bias, as well as sighting probability bias (6).

Due to the combined effects of visibility bias and sighting probability bias, species such as black rhinoceros, and to a lesser extent kudu, are impossible to count from the air. Other less conspicuous species are also likely to be underestimated, even when applying the double-count method. As a rule of thumb, aerial total counts should only be used for highly visible species. However, aerial total counts may be the only option in situations where clumping or topography prevent sampling.

Most of the errors and biases discussed here also apply to aerial sample counts that will be discussed in the next chapter.

5.2 Census Design

When designing an aerial census, the following steps should be taken:

1. Take care in selecting the most appropriate season for the count. The peak dry season gives results that are less affected by visibility bias and sighting probability bias. Although during the wet season there is less dust in the air, the abundant foliage renders certain species invisible. However, during the peak dry season, certain conservation areas lack water, while parts may be burnt. This often causes animals to concentrate near the remaining water sources and in unburnt areas. Under these conditions, the early dry season may be a more appropriate period to carry out a census.

2. Obtain a good map of the census zone, preferably at a scale of 1:250,000.

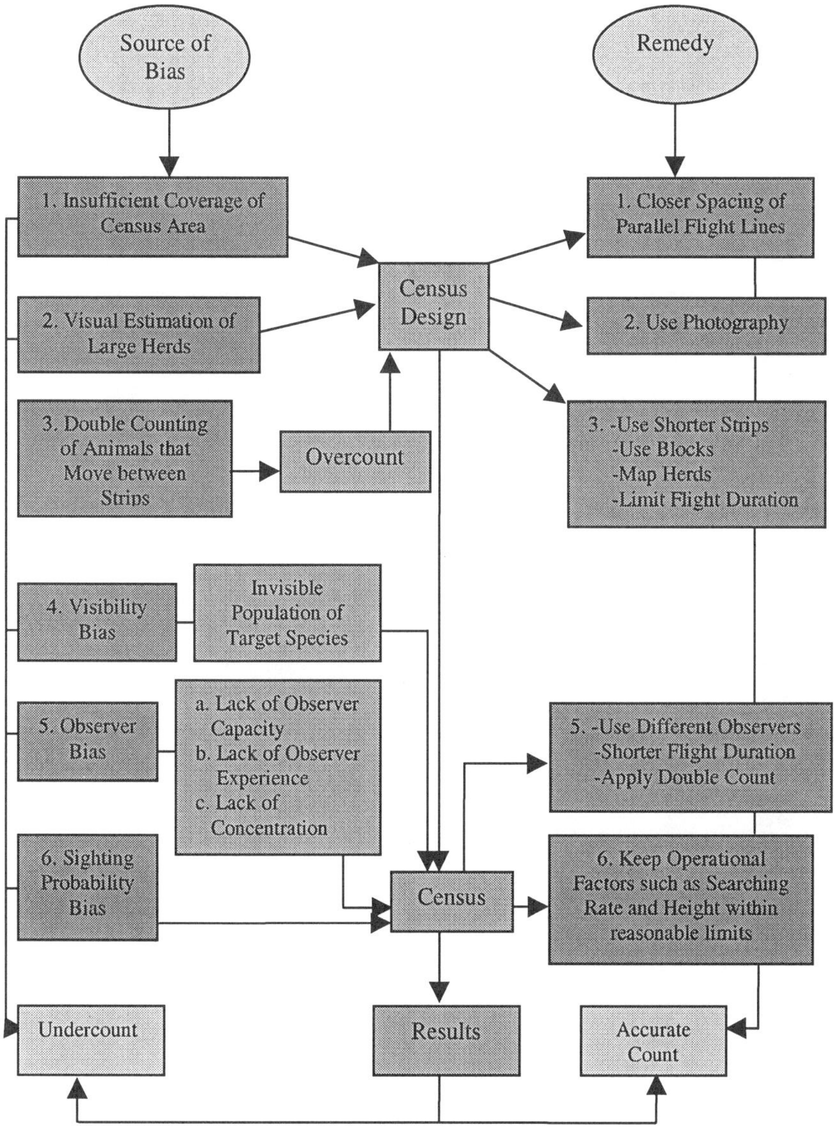

Figure 5.1: Diagrammatic representation of sources of error and bias in aerial census, and the possible remedies to improve the final estimates of population abundance.

3. Divide the census zone into discrete blocks that can be covered by one aircraft in a single day. Remember that a maximum of three 2-hour counting sessions can be flown in a day, not including placement time. Counting blocks may vary in size, but should not exceed approximately 1,000 km^2 and should be defined by features such as roads, escarpments, mountain ranges or water bodies (Figure 5.2).

4. Place flight lines in each block at intervals of between 500 m and 2 km, depending on the type of vegetation, the topography and the target species. Counting elephants in savanna woodland may require spacing of 500 m, whereas intervals of 2 km may be used when counting elephants in large open grasslands. Here, we should keep in mind that short flight lines may be spaced close together. However, long flight lines should not be closely spaced so as to avoid double counting.

5. Take care to maintain the specified altitude and speed. One hundred metres is usually the most appropriate altitude, depending upon the target species and the habitat. As a rule of thumb, searching rates should be kept within the following limits:

   $<$ 250 km^2 per hour for wide open areas,

   $<$ 150 km^2 per hour for open savanna woodland, and

   $<$ 100 km^2 per hour for dense savanna woodland.

6. Use a GPS to plot the course of the aircraft and to maintain straight flight lines. If possible, record waypoints of each observation, but always use the time to show the position of the animals observed along the flight line.

7. Take photographs of herds with more than 20 individuals, and also record the number of animals counted in the herd. When photographing large herds, fly a straight line alongside the herd, instead of circling it. Circling makes it difficult to work out the overlap of the photographs covering the herd (Norton-Griffiths, 1978).

8. In woodland areas, and where the target species lives in small groups, use the double-count method to correct for observer bias

and sighting probability bias (and a small part of visibility bias). When using the double-count procedure, make sure that all observers carry a watch that is synchronised before taking off. For more details on the double-count method *see section 5.4* below.

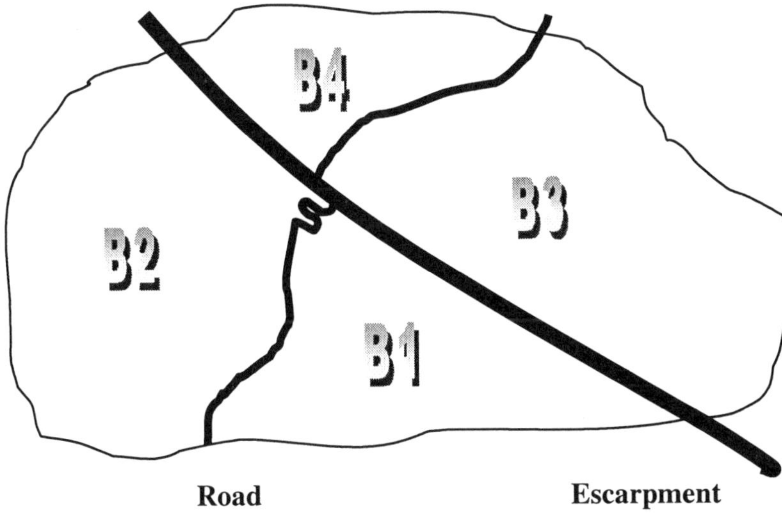

Figure 5.2: Hypothetical example of a conservation area with an escarpment running from south-east to north-west, and a single road running from south-west to north-east. These features allow the area to be divided into 4 counting blocks B1 to B4, with flight lines running parallel to the escarpment.

9. Use pre-printed data sheets to record sightings and other relevant information (Table 5.1).

10. Count a pre-selected sample of blocks repeatedly to assess the accuracy of the census. These blocks should be representative of the entire census zone. Bias towards undercounting may be in the range of 10% to 50%, depending upon the species and the type of habitat, but it will be reduced by applying the double-count method.

Table 5.1: Example of a data sheet that can be used for an aerial census.

<table>
<tr><td colspan="2">Pilot:</td><td colspan="2">Navigator:</td><td colspan="3">Date:</td></tr>
<tr><td colspan="2">Mid-left:</td><td colspan="2">Mid-right:</td><td colspan="3">Conservation Area:</td></tr>
<tr><td colspan="2">Rear-left:</td><td colspan="2">Rear-right:</td><td colspan="3">Block Number:</td></tr>
<tr><td colspan="2">Time Start:</td><td colspan="2">Time Finish:</td><td colspan="3">Flight Line Number:</td></tr>
<tr><td>Time</td><td>Species</td><td>Estimate</td><td>Distance</td><td>Photos</td><td>#</td><td>Remarks/ Waypoint</td></tr>
<tr><td></td><td></td><td></td><td></td><td></td><td></td><td></td></tr>
<tr><td></td><td></td><td></td><td></td><td></td><td></td><td></td></tr>
<tr><td></td><td></td><td></td><td></td><td></td><td></td><td></td></tr>
<tr><td></td><td></td><td></td><td></td><td></td><td></td><td></td></tr>
</table>

5.3 Aerial Total Counts of Hippos and Crocodiles

Aerial counting of crocodiles is only feasible during the peak dry season, when the rivers are low and the water is clear. A slow and low flying aircraft should be used, such as a micro-light. In the central Luangwa Valley, crocodiles were seen from a height of about 50 m, and densities of 12 crocodiles per km of river frontage could easily be counted at speeds between 50 and 60 knots (93 – 112 km/hr). The resulting estimates may be low by about 5%, due to small crocodiles missed during the count.

A micro-light, flying at the same speed and height as for crocodile counts, is also the most suitable aircraft for hippo censuses. The census should be conducted during the peak dry season, when hippo congregate in the deeper pools. If a larger and faster aircraft, such as a Cessna 206, is used, then large schools of hippos should be photographed (Figure 5.3). Single hippos and small groups should be recorded on paper, while large schools can be counted from the photographs (Figure 5.3). The resulting estimate may be low by about 10%, due to small calves and submerged hippos missed during the count.

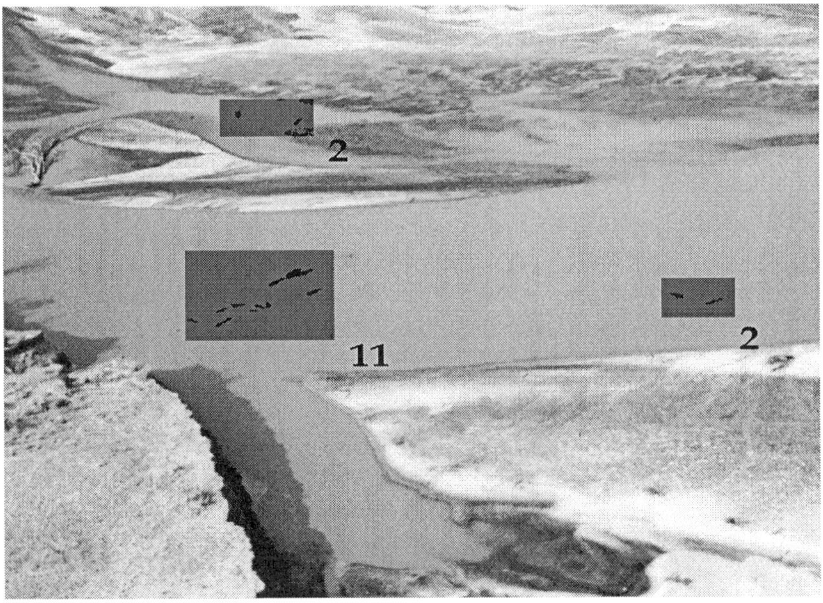

Figure 5.3: Photograph of hippo in the Luangwa River. Fifteen hippo were counted in this frame.

5.4 The Double Count Method

Over the past 25 years researchers have been intrigued by the problems of observer bias and sighting probability bias in aerial sample surveys. The same problems are inherent in aerial total counts. The first approach to solving this problem is to carry out a ground count of a representative part of the survey area that provides a better estimate than the corresponding aerial count. The ground count can then be used to correct the aerial count for negative bias (Jolly and Watson, 1979). However, as we have seen in the previous chapters, accurate ground counts are rarely feasible and therefore this approach has limited application.

A second approach is to use multiple linear regression analysis to model the effects of height above ground level, strip width and speed of the aircraft on the trend of observed density. A correction factor can then be estimated by extrapolating the observed range of variables to zero values (Caughley, 1974; Caughley *et al.*, 1976). This approach requires a lot of additional flying, making it an expensive exercise and therefore not very suitable to the African situation. In addition, the approach has been criticised for using linear regression analysis and extrapolation to zero values (Barnes *et al.*, 1986). Furthermore, it has been shown that in certain cases the relationship between sightability and height is non-linear when approaching zero values (Norton-Griffiths and Santibanez, 1980).

The only approach with a wide application in the field is the double-count technique. This technique uses the concept of the mark/recapture model. Its potential use for estimating observer bias in aerial survey was introduced by Caughley (1974). Since then, different adaptations have been applied in a wide range of circumstances for a number of species, ranging from bald eagles in Ontario (Grier *et al.*, 1981) to emus, feral horses and donkeys in Australia (Caughley and Grice, 1982; Graham and Bell, 1989). Most applications use an adaptation of the Petersen estimate (Seber, 1982):

$$\acute{Y} = y_1 \times y_2/m$$

For the mark/recapture method:

y_1 is the number of marked animals,
y_2 is the number of animals caught,
m is the number of marked animals recaptured.

Similarly, for the double-count technique:

y_1 is the number of animals seen by the front observer,
y_2 is the number of animals seen by the rear observer,
m is the number of animals seen by both observers.

The double-count techniques developed in Australia can be applied to aerial counts of African large ungulates (Caughley and Grice, 1982; Graham and Bell, 1989). In the double counts conducted by Caughley and Grice (1982), two observers sitting in line made simultaneous counts of emus. The two observers acted independently and without collusion. The density of emus was so low that the observers only recorded the absence or presence of emus in 1-km^2 units. Therefore, if both tandem observers recorded emus on a given unit, then only one group was expected to be present and it was assumed that both of them saw the same group. With high animal densities, however, this method poses a problem with regard to the identities of the animals seen by both observers. The double-count method introduced by Graham and Bell (1989) was supposed to solve this problem. During their double counts, two observers were sitting in line to make simultaneous counts of feral horses and donkeys. Both observers followed identical counting procedures, differing only in their recording duties. The rear observer had to notify the front observer of every observation he made. The front observer would then record the group as seen by him also or by the rear observer only. At the same time, the front observer would establish a third category for his observations only. These data were used for estimating observer bias (Graham and Bell, 1989).

The problem with the first approach is that it will not work in situations with high animal densities, while the second approach may generate biased observations due to communication between the two observers. This can be easily explained in psychological terms. Imagine a situation whereby one of two observers counting in tandem has poorer eye sight and/or lesser experience than the other. The first observer may not like to admit his deficiencies and his cheating may bias the results.

In the Luangwa Valley we partly resolved this problem by having two observers sitting in line, counting independently and without collusion, but recording the exact time of each observation. Communication between observers during flight was not allowed. Prior to a survey, all watches were synchronised. When two observers sitting in line made an observation of the same species at exactly the same moment, we considered this to be the same group of animals. When the observations differed in group size, it simply meant that one observer counted the animals more accurately. Frequently, two teams of observers were used to improve the accuracy of the correction factor.

By comparing the observations and the exact times of each observation of the front and rear observers, the data can be grouped into three categories (Caughley and Grice, 1982):

S_1, the number of animals seen by the first observer but missed by the second,
S_2, the number of animals seen by the second observer but missed by the first,
B, the number of animals seen by both observers.

If both observers counted the same group of animals, but arrived at a different estimate of group size, then the lowest estimate was used for category B, whereas the difference between the two estimates was either used for category S_1 or S_2, depending on which observer had the highest estimate.

The probability that animals will be seen by both observers is (Graham and Bell,1989):

$$P = \frac{B(S_1 + S_2 + B)}{(S_1 + B)(S_2 + B)}$$

where the variance of P can be estimated as:

$$\text{var}(P) = \frac{S_1 S_2 (S_1 + S_2) P}{(S_1 + B)^2 (S_2 + B)^2}$$

The observer bias correction factor (C) is 1/P.

The uncorrected population estimate is calculated using the data recorded by the front observer(s). It is then multiplied by the bias correction factor C to give the corrected population estimate.

The variance in the final population estimate, corrected for observer bias, can be approximated by multiplying the variance of the uncorrected population estimate by the correction factor C.

When using the double-count method, one important assumption and several important conditions must be considered. The assumption, central to all mark/recapture methods, is that animals are equally catchable, or in terms of aerial counts, that all animals in the strip have an equal chance of being seen by a given observer. Firstly, this assumption is false, because large groups are easier to detect than small groups, animals on the move are easier to detect than stationary groups, and so on. Secondly, this is circular reasoning, because observers differ in eye sight, capabilities to concentrate and levels of experience. Moreover, the whole exercise is set up to correct the results for these differences in observer capabilities.

Probabilities of detection will be distributed between 0 and 1 and may follow a beta distribution (Caughley and Grice, 1982). Violation of the assumption of equal catchability does not result automatically in a biased estimate of population size, because both the probability of an animal being marked and that of its recapture are random draws from a beta distribution (Magnusson *et al.*, 1978). However, simulations showed that at low sightabilities (less than 45% of the population is observed), under-estimations of correction factors by as much as 10% may occur (Caughley and Grice, 1982). Thus, the double-count method should preferably not be applied in conditions where the density of the vegetation obstructs visibility and in situations where the target species are not very conspicuous.

When applying the double-count procedure, one should remember that a particular correction factor is only valid for that particular census or survey and that species. Even when the count is repeated with the same aircraft, the same crew and observers, the same design, counting the same species at the same time of the year, it will still differ from the first one. For instance, the aerial counts in the Luangwa Valley were usually conducted during October. However, every wet season has a different rainfall pattern, resulting in annual differences in water availability and vegetation density in October. Most of the other relevant factors, such as pilot and observer capabilities will also show minor changes from year to year. A good example is the enormous variation in observer bias correction factors generated in the Luangwa Valley between 1993 and 1998 (Table 5.2). With the exception of the 1999 survey (July), the surveys were flown in October. However, different teams of experienced observers were used each year. The survey design remained constant up to 1994, after which the altitude was raised from 100 m to 150 m, and the strip width increased from about 140 m to 200 m. In addition, during the 1998 survey, two out of four observers were inexperienced. The correction factors for each species vary from year to year, while those for 1998 are substantially higher (Table 5.2). The higher values in 1998 were caused mainly by changes in altitude and strip width and the two inexperienced observers.

Table 5.2: Correction factors for 12 large ungulate species counted in the Luangwa Valley between 1993 and 1999, generated by applying the double-count method.

SPECIES	CORRECTION FACTORS				
	1993a	1993b	1994	1998	1999
Buffalo	1.14	1.41	1.20	2.30	1.12
Eland	1.00	1.60	1.00	2.14	-
Elephant	1.06	1.27	1.39	1.88	1.15
Hartebeest	2.06	3.08	2.00	-	1.51
Impala	-	-	1.26	2.09	2.33
Kudu	1.00	2.25	1.57	2.16	1.70
Puku	-	-	2.14	1.47	1.80
Roan	1.67	1.67	4.00	-	1.90
Warthog	-	-	2.47	3.83	2.50
Waterbuck	1.17	1.50	2.00	2.10	3.45
Wildebeest	-	-	1.12	-	1.10
Zebra	1.18	1.41	1.41	2.54	1.35

5.5 Field Example of Observer Bias Experiments

In 1993, two aerial sample counts were carried out in the SLAMU area (South Luangwa Area Management Unit, formerly Luangwa Integrated Resource Development Project or LIRDP). The first count was conducted in the early dry season (4-7 June) and covered the entire project area, e.g. the South Luangwa National Park (SLNP) and the Lupande GMA. Only elephants were counted. Transects were flown accurately with the aid of a GPS, whereby the beginning and the end of a transect were determined by distinct features in the landscape. In SLNP, transects were flown east to west between the Luangwa River and the Muchinga escarpment. In the Lower Lupande GMA transects were flown east to west between the Luangwa and Lupande Rivers, and in the Upper Lupande GMA, transects were flown east to west between the Luangwa River and the eastern boundary of the GMA (Figure 5.4). The second count was carried out in the late dry season (16-18 October). It covered only the SLNP, and nine species were counted, including elephants. For both surveys, flight paths were 5 km apart (Figure 5.4), with a sampling intensity of 6%.

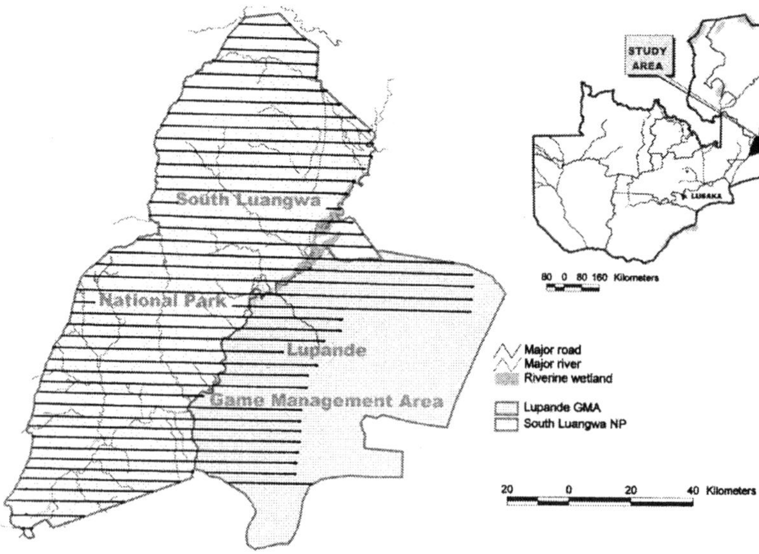

Figure 5.4: Location of the SLAMU area in Zambia (above); the South Luangwa National Park and the Lupande Game Management Area, with the Luangwa River intersecting the park and the GMA (below). In 1993, flight paths were 5 km apart. However, in surveys done after 1993, flight paths were 2 minutes apart.

A Cessna 206 aircraft flew at 100m above ground level (AGL). A barometric altimeter was used for height control, but it did not work as well in the hilly parts of the Lupande GMA as in the mainly flat parts of SLNP. However, wildlife densities in the hilly areas of the GMA were extremely low, limiting this potential source of error. The strip widths were calibrated by moving the streamers until the average strip width was approximately 280 m. Average flying speed was 110 knots, giving a searching rate of 57km²/hour.

During the first survey, the aircraft was manned by a pilot, a navigator and two experienced observers. During the second survey, however, two additional observers counted in the back of the aircraft. A total of 12 observers was available, each with a different level of experience. The double-count procedure was followed to permit correction for sighting probability bias and to estimate the bias related to observer experience (*see section 5.4*).

For each session, the number of visible animals in the transects was calculated using the Petersen Estimate (*see section 5.4*). Then, for each observer, the percentage of animals seen was calculated. With the percentage of animals seen by each observer for each separate session as the dependent variable, a multiple-linear regression was performed, using total aerial survey experience (hours), current survey experience (hours spent counting in this particular survey) and flight duration (session in hours) as the independent variables. The results are summarised in Table 5.3. Tests of the multiple-regression assumptions were performed and met in all cases. The resulting regression coefficients were compared and tested with t-statistics to determine which of the independent variables has the most influence on the relationship. Final analysis of population estimates followed Jolly's Method 2 for unequal sized sampling units (Norton-Griffiths, 1978), which will be discussed in the next chapter.

Although the model provided a reasonable fit, with a multiple correlation coefficient of 0.704 and a significant F-statistic ($P<0.05$), it explained only 50% of the variation. This implies that several other variables that were not tested, such as eye-sight of the observer, capability to concentrate for long periods, familiarity with the area, cloud cover and sunlight also play an important role. Both total survey experience (T) and current survey experience (C) had significant positive coefficients, while flight duration (F) was almost significant at the 5% level, but with a negative coefficient (Table 5.3). The overall equation has the following form (Jachmann, 1995a):

$$\% \text{ Animals Observed} = 0.28\,T + 4.01\,C - 7.68F + 77.49$$

Thus, with increasing total survey experience and current survey experience more animals were observed, whereas during long flights fewer animals were observed. The intuitive explanation for this is that each observer has to form a series of searching patterns, a process that depends upon the capabilities of the observer, the abundance of the species, the state of the vegetation and the number of species to be counted. With an increasing experience, this process may develop more rapidly. Also, it follows that single-species counts will produce better results than multi-species counts. Trying to count too many species at once is a common mistake.

Table 5.3: Estimated regression coefficients, Standard Errors (SE), percentage Standard Error (% SE), computed t-values (t) and significance levels (p). The dependent variable is the percentage of animals of the visible population seen by each observer.

Variable	Coefficient	SE	%SE	t	p
Total Survey Experience (T)	0.279	0.105	37.73	2.650	0.02
Current survey Experience (C)	4.006	1.559	38.92	2.569	0.02
Flight Duration (F)	-7.680	4.678	60.92	-1.642	N.S.
Intercept	77.488				

Using the above formula, we can estimate the optimum flight duration for a single session, as well as the level of experience required to observe 95% of the visible population in the survey area. The most experienced observer during the October 1993 survey had a total survey experience of approximately 100 hours. Using a current survey experience of one hour, the optimum flight duration for this particular observer would be about two hours. Any flight duration longer than two hours would reduce his efficiency. This implies that a less experienced observer would require a longer current survey experience, i.e. a longer session to get acquainted with the area and target species, or alternatively shorter flights. However, flights shorter than two hours are uneconomical.

The survey in June 1993 gave an estimate of 5,263 ± 1,081 elephants for SLNP. The count in October 1993 gave an estimate of 5,270 ± 680 elephants for SLNP. Although the second count used an observer bias correction factor, the two different survey designs gave identical results. The double-count method resulted in an additional 6% of elephants observed. A correction factor of only 6% is negligible compared with the % Standard Errors of 20.5% (June) and 12.9% (October). Hence, in this case, the slight gain in precision and accuracy with the double-count procedure did not outweigh the extra fuel expenses and the risks involved in low-level flying with six people on board. In contrast, observer bias correction factors were substantially higher for the other 8 species and significantly contributed to the accuracy of the final population estimates.

5.6 Field Example of Visibility Bias and Sighting Probability Bias Experiment

The double-count technique will compensate for only some of the animals missed through sighting probability bias, while an unknown number of animals will be missed due to visibility bias. Numbers missed depend on the density of the vegetation and the conspicuousness of the species (e.g. body size, colour pattern and group size). With large conspicuous species such as elephant and buffalo, accumulated biases give underestimates of 10% to 15% on average. Aerial counts of solitary species, such as duiker and grysbok, return fractions of the true numbers of animals. Thus, aerial techniques are not suitable for small solitary species. But how reliable are aerial techniques when counting the slightly larger gregarious antelopes, such as puku? In other words, how many animals are missed due to visibility bias and sighting probability bias that are not corrected for by applying a double count?

In an attempt to answer this question, numbers of puku in Kasanka National Park were assessed with an aerial total count, an aerial sample count and a total ground count in October 1999 (Jachmann and Phiri, 1999a). The small Kasanka National Park (470 km^2) is located in the north of Zambia, and is covered by miombo woodland (66%), dominated by *Brachystegia*, *Julbernardia* and *Isoberlinea* species, with large open dambos (16%), and small areas of permanent swamp and dry evergreen forest. Puku are the commonest animals in the park. During most of the day, puku are found in the open dambos, where visibility is excellent, both from the ground and from the air.

For the aerial total count the park was divided into 10 blocks that covered all of the dambos known to contain puku. The census was flown with an average airspeed of approximately 100 knots, at 50 m above ground level. The aircraft (Cessna 206) followed the rivers, while two observers counted puku to the left and to the right, with the river being the strip below the aircraft. The aerial sample count was done the following day, with a sampling intensity of 42.1%. A total of 27 transects were flown east – west on 30 sec. grid lines. Calibrated transect width was 337.2 m. Two front observers counted puku to the left and to the right of the aircraft, with an additional observer in the back for the double count. Two days later, a total ground count was done, whereby teams of 2 scouts, equipped with binoculars, counted every dambo in the park.

Table 5.4: Results of aerial and ground counts of puku, conducted in Kasanka National Park in October 1999. PE is Population Estimate, and SE is Standard Error, obtained from repeat counts of the same blocks.

METHOD								
AERIAL SAMPLE COUNT			AERIAL TOTAL COUNT			TOTAL GROUND COUNT		
PE	SE	95% CL	PE	SE	95% CL	PE	SE	95% CL
1,392	393	622–2,126	2,884	191	2,464–3,304	3,660	366	2,943-4,377

The observer bias correction factor was 1.91, implying that almost half of the pukus were missed by the front observers. The total ground count returned 3,660 puku, with an estimated standard error of approximately 10% (Table 5.4). The aerial sample count underestimated puku by 62%, and the aerial total count by 21% (Table 5.4). Thus, in spite of a double count to correct for observer bias, the combined effects of visibility bias and sighting probability bias resulted in an undercount by as much as 62% for the aerial sample count.

What can we learn from this? Aerial counts of puku return underestimates even under conditions of excellent visibility. Thus, when accurate estimates of population abundance are required, medium-sized gregarious antelopes should not be counted from the air. Instead, ground count techniques should be used. With large study areas, where ground counts are too expensive, results from aerial counts should be considered conservative estimates, and interpreted with caution.

Chapter 6

AERIAL SAMPLE COUNTS

Aerial sample counting is the most frequently used method to assess the abundance of large ungulate species on the African continent. Since its first use in the 1950s, the technique has changed in many aspects. The accuracy has improved beyond recognition, mainly as a result of the introduction of the GPS as an aid in navigation, and a better understanding of the potential sources of bias. The principles of an aerial sample count are the same as those that apply to other methods of sample counting discussed in previous chapters. As with other sample techniques, aerial sample counts estimate the number of animals by counting in a small part of the survey area. This is known as the sample. The density estimate for the sample is then extrapolated to the whole survey area. The calculations are similar to those for sample counts on the ground.

6.1 Choice of Sample Unit

In aerial sample counting three types of sample units are recognised: quadrats, blocks and transects (also known as strips). Procedures for counting quadrats and blocks are similar, with the only distinction that a quadrat refers to a square or rectangular area, whereas a block may be an irregular shaped area defined by features on the ground (Norton-Griffiths, 1978). When quadrats are the sample unit, the survey area is divided into grid squares or rectangular areas of appropriate size. Some of these units are then selected to form the sample. Units may be chosen randomly, or by following a systematic pattern. Each of the selected units is counted. The main problem with using quadrats is the lack of physical marks by which the observers can judge whether or not a group of animals falls within the unit. This boundary problem can lead to major errors in the final results that cannot be corrected. Thus, it is strongly advised not to use quadrats.

6.1.1 Blocks

In block sampling, units are demarcated by physical features on the ground, such as roads, tracks, hill ranges, escarpments, distinct habitat features such as a transition from woodland to grassland, and drainage systems. As with quadrats, blocks can be selected randomly, or placed in a systematic pattern. Procedures are exactly the same as for aerial total counts: an attempt is made to count each animal of the target species within the selected blocks. Due to the clear demarcation of block boundaries, it is easy to decide whether or not animals fall within the boundaries of the sample unit. However, there are several disadvantages to using blocks.

1. Blocks return poor information on the spatial distribution of the target species.

2. Errors may be caused by the target species concentrating near the features used as boundaries, such as rivers or streams. Animals may move in and out of the block while counting is in progress (Norton-Griffiths, 1978).

3. When the target species is unevenly distributed over the survey area, some blocks may contain large numbers of animals and others none. This leads to a large variance between blocks and therefore an imprecise estimate of abundance.

> As an example we take a hypothetical survey area, traversed by 3 roads, with a lechwe population of unknown size and distribution, but which appears to be concentrated in 3 large herds. Funds are available to carry out an aerial sample survey to estimate the size of the lechwe population and to assess its distribution over the survey area. Using the roads as boundaries, the area is divided into 6 blocks (Figure 6.1). A systematic block sample count, sampling every other block and starting with block 2, will result in 3 blocks with few lechwe (Figure 6.1). The estimate would therefore be inaccurate and imprecise. If we would start the count with block 1, our sample of 3 blocks would contain most of the lechwe population. The resulting estimate would be relatively accurate, but still imprecise. A sound alternative would be to divide the survey area into smaller blocks. When many small units are counted rather than a few large ones, the resulting estimate will be more precise. The main disadvantages of this alternative are the increased costs of the survey, and the problem of delineating certain blocks without the presence of distinct features in the landscape. This will create a boundary problem, as is the case with quadrat sampling. In this particular example, a total count of the lechwe population, using photography, would be the ideal alternative.

However, without any prior knowledge about the size and the distribution of the population, and funds that are sufficient for a sample count only, the alternative would be to count the lechwe using transects, placed at regular intervals. In our example, 7 out of 8 parallel transects contain lechwe, considerably improving the precision of the estimate and returning adequate information on the distribution of the lechwe population (Figure 6.1).

4. In block counts flight paths are irregular and placement time (time taken to fly from one unit to the next) is high, making block counts inefficient in terms of ground covered per unit effort. Consequently, block counts are usually more expensive than transect counts. Smaller blocks may give a more accurate and precise estimate but increase the time spent flying between blocks and so increase the cost.

5. Observer fatigue is more common during block counts than during transect counts, due to irregular flight paths, with a lot of steep turns and backtracking. Therefore, observer bias and sighting probability bias may be more important in block counts than in transect counts.

When should we use blocks as the sampling unit? In hilly or mountainous terrain poor height control results in a variable transect width and inaccurate estimates of abundance. The use of blocks is advised under such conditions. Survey areas with a dense vegetation may require a searching pattern with irregular flight paths and variable heights above ground level to flush out animals. Blocks may be the appropriate choice under these conditions.

In conclusion, block counts are useful when transect counts are not feasible, such as under conditions of mountainous and broken terrain and high vegetation density. In all other cases transect counts are superior to block counts.

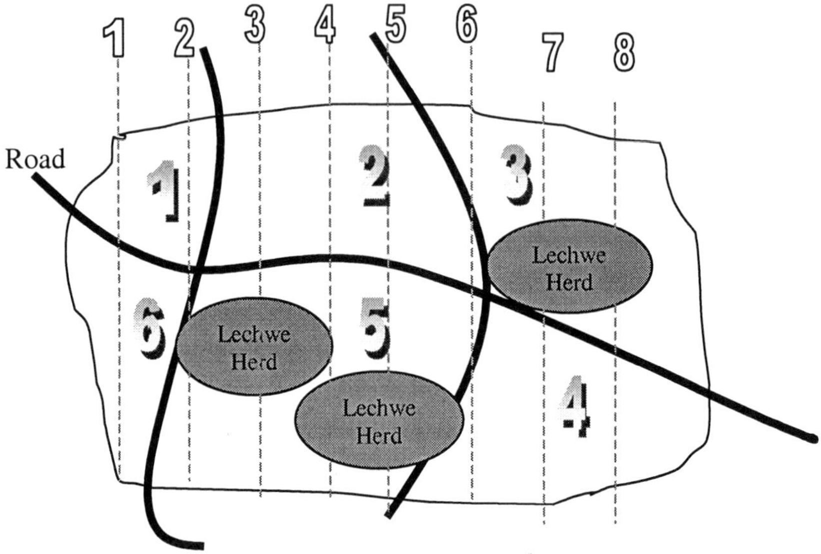

Figure 6.1: Hypothetical survey area, divided into 6 blocks using the road system (thick lines) as boundaries. The objective of the survey is to count the lechwe population that occurs in three large herds (grey areas). Eight parallel transects for an aerial survey have been indicated (broken lines).

6.1.2 Transects

Aerial transect counting is the most efficient and usually the most appropriate aerial sampling technique. The basic principle is simple: the aircraft follows a straight flight path from one side of the survey area to the other, at a fixed height above ground level and at a fixed speed. Streamers, ropes with funnels on the end, are attached to the wing struts of the aircraft to demarcate a transect or strip on the ground. The width of the strip is part of the survey design, and should be decided prior to the survey. The altitude at which the survey will be flown should also be decided in advance. The streamers should be positioned to give the desired strip width at that height. During the survey, the observers count all animals of the target species in the strips.

Generally, transect or strip sampling is more flexible than block sampling, with the option of altering the size of the unit by simply altering the strip width. However, transect sampling suffers from more potential sources of bias than block sampling. In transect sampling observers have only one chance to count groups of animals, on one hand minimising bias due to double counting, but on the other hand increasing visibility bias and sighting probability bias. However, observer fatigue during block sampling may also lead to increased visibility bias and sighting probability bias. In summary, transect sampling is superior to block sampling in terms of cost-efficiency, precision and bias due to double counting.

6.2 Survey Design

The survey design and field procedures for block sampling follow those for aerial total counts. The only difference from an aerial total count is that in block sampling only some of the blocks are counted. Therefore, we will focus on the survey design for transect sampling. When designing an aerial survey the following steps should be followed:

1. **Define the objective**: Which species are to be counted in which habitat, and for what reasons? This determines what levels of precision will be acceptable, and whether the budget will cover the expenses.

2. **Define the survey area**: A good map of the area should be obtained, preferably with a scale of between 1 : 100,000 and 1 : 250,000 (1 cm on the map represents 100,000 cm or 250,000 cm on the ground). In Zambia, most maps use a cylinder projection such as the Transverse Mercator. A map made with a cylinder projection has parallels (running parallel to the equator) and meridians (circles running through the North and South Poles) at right angles. These lines are not equidistant, which means that one cannot measure the distance with a ruler. Instead, minutes and seconds latitude will have to be used to measure distances (1 minute latitude is 1,852 m, which is one Nautical Mile). The next step is to determine the boundaries of the survey area. This should be straightforward if the survey area is a protected area with resident animal populations, or surrounded by cultivation that limits animal movement across boundaries. If, however, the survey area is part of a larger wilderness where animals can move freely, setting boundaries depends upon the objective of the survey. If the objective is to estimate the number of animals in a particular area at a particular time of the year, then the boundaries are those of the area in question. If the objective is to estimate the size of a particular population, then the boundaries are defined by the area occupied by that population. Here, we require detailed information on the spatial and temporal distribution of the target

species prior to the survey. Less straightforward is a situation where the objective of the survey is to estimate accurately the abundance of migratory animal populations in an area that forms part of a vast wilderness. This situation occurs in the Luangwa Valley, where hunting quotas are set for a variety of species for a large number of GMAs and Open Areas that together with 4 National Parks form a continuous wilderness of approximately 100,000 km². Due to limited law-enforcement operations in particular parts of the Luangwa Valley, most migratory species form distinct populations, but with some overlap into adjacent areas. For the purpose of setting hunting quotas, these adjacent areas are included in the annual aerial surveys.

3. **Determine whether stratification is required**: For making decisions about stratification one needs information beforehand on the temporal and the spatial distribution and the local densities of the species in question. Stratification of the survey zone becomes a useful tool to improve the precision of the estimate when the difference in density between the various strata is at least three times. As discussed previously, it is difficult to achieve one stratification for several species that are counted during the same survey. Use the formula in section 2.4 to optimise sample allocation for the various strata. Also, stratification may be used to divide a survey area into units of a manageable size. In the Luangwa Valley, our experience has been that the size of the South Luangwa National Park (9,050 km²) is the largest area that can be treated as a single survey zone. Any areas larger than this should be divided into smaller units and sampled as separate strata.

   Stratification after the survey is sometimes done with multi-species counts, to obtain a more or less optimal stratification for each species counted. This procedure is questionable from the statistical point of view and does not incorporate optimal sampling allocation, but may improve the precision of the estimates.

4. **Establish a base-line**: A base-line should be drawn on the map such that transects placed at right angles to the base-line cover the entire survey area (Figure 6.2). Transects must be oriented so that they run along a density gradient of the species in question (Figure 6.2). If transects are located across a density gradient

(east-west in Figure 6.2), some will contain many animals and others none, resulting in an estimate with low precision. If transects are located along a density gradient (north-south in Figure 6.2), most of them will contain animals, reducing the sampling error. If there exists no information on the distribution, transects should run at right angles to the drainage system or any other features where animals tend to concentrate. Once the base-line has been established, it is divided into sections of the same width as the selected transect width. With transects of 250 m on both sides of the aircraft, a 100 km base-line would be divided into 100/0.5 = 200 units. This is the maximum number of units from which the sample can be drawn.

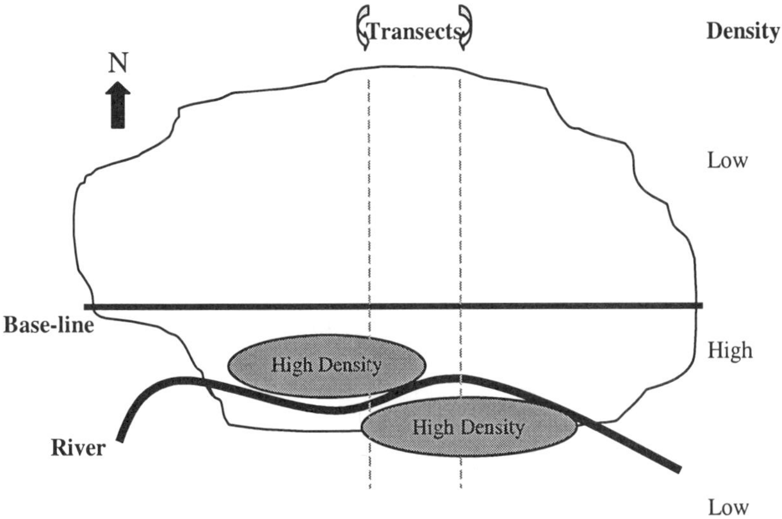

Figure 6.2: Illustrating a base-line that allows transects to cover the whole area of the survey zone. Transects run along the density gradient of the target species.

5. **Random or systematic sampling**: Although we have discussed this in section 2.5, we would like to make the following comments: Most of the alleged differences in the two methods are imaginary rather than real, except for one; the main advantage of systematic sampling is that it returns better information on temporal and spatial distribution. In addition, it is also easier to carry out in practice.

6. **Determine the sample effort**: After the base-line has been established and a decision has been made on the sampling method, the sampling intensity or area coverage will have to be

determined. The sampling intensity is a function of the number of transects in the sample and the selected transect width. With a wider strip width and more transects, the intensity of the survey will increase. The choice of sampling intensity depends on the objective of the count, the budget and the density and distribution of the target species. The width of the transect must be selected to maximise area coverage, but to minimise bias.

> The effect of transect width on sightability was investigated for a series of elephant counts conducted in the South Luangwa National Park, between 1965 and 1973 (Caughley, 1974). Without going into details, the general conclusion was that with an increasing transect width, sightability of elephants steeply declined. Similar experiments were conducted in Kenya with wildebeest and multi-species counts. As transect width is increased, the time available to locate, recognise and count animals decreases, and visibility and sighting probability biases increase. The optimum transect width is between 100 m and 200 m (Caughley, 1974).

7. **Placement of transects**: Whether the transects are placed at random or systematically, the general rule is that the larger the sample size, the lower the sampling error. The objective of the count determines the required precision, which determines the sample size or the number of units that should be counted. The relationship between sample size, density and precision will be discussed in detail below (*see section 6.4*). When a decision has been made on the sample size, transects should be placed at right angles to the base-line. The beginning and the end of each transect should be indicated with latitude and longitude co-ordinates to simplify navigation with a GPS. The invention of the GPS has made accurate navigation much easier and we advise using minutes and seconds latitude or longitude to indicate the distance between transects.

8. **Choice of aircraft**: The best type of aircraft for survey work is a high wing monoplane with wing struts to attach the streamers. The aircraft should have at least 4 seats to enable double counting. The engine should be powerful enough to carry 4 people and a large fuel load, and have reserve power to manoeuvre in difficult situations. The most suitable aircraft is the Cessna 206 with room for 6 passengers. A similar aircraft, the Cessna 210, is designed for higher speed and therefore has no wing struts. This makes it

unsuitable for survey work. Other commonly used aircraft types are the Cessna 180, 182 and 185.

9. **Set flight parameters**: The correct height above ground level and flying speed will minimise bias. Both height and speed have a significant effect on sightability (Caughley, 1974). Safety is an important consideration, especially with low-altitude flying. Height should not be lower than 100 m, but also not more than 200 m. Bear in mind the lowest safe operating speed for the type of aircraft. The speed of the aircraft multiplied by the transect width gives the searching rate. Ideally, the searching rate should be between 75 and 100 km^2 per hour. With a transect width of 2 x 200 m (0.4 km), this would give a speed of between 187.5 and 250 km per hour, corresponding to 101 to 135 knots or nautical miles per hour.

10. **Flying the survey**: Navigation is straightforward using a GPS. Height control is achieved with a radar altimeter. Prior to a flight, the radar altimeter has to be calibrated against the pressure altimeter. The aircraft is flown a few meters above ground level and the pressure altimeter is set at zero. The aircraft is then flown to the height at which the survey will be conducted. The next step is to make a few passes over the location where the pressure altimeter was set to record the readings from both altimeters. The radar altimeter can then be calibrated against the pressure altimeter. Height control is never perfect and tends to follow a sine pattern when flying the survey. To make corrections for the actual height flown, readings from the radar altimeter should be recorded by the navigator every 5 minutes. Later, the actual transect width can be calculated using the mean height above ground level given by the figures and the measurements in Figure 6.3.

6.3 Positioning the Streamers

For a tail-wheel aircraft, the tail wheel should be lifted onto a block, so the aircraft is in the flying position. The streamers are positioned as follows (after Norton-Griffiths, 1978):

The observers should be sitting in the parked aircraft in a comfortable and relaxed position.

1. Measure the height of the observer's eye (h) above ground (Figure 6.3),

2. Fix the position of the inner streamer (i) as low as possible on the wing strut, ensuring that the line of sight is clear of any aircraft parts (Figure 6.3). Place a marker on the ground on this line of sight (A), while the observer uses a small piece of tape to mark the window (a), creating a line of sight that passes from a to i to A.

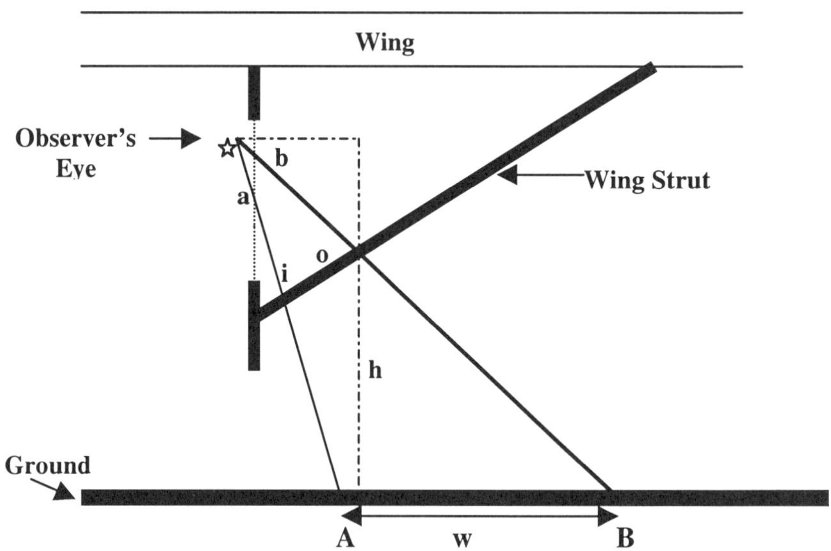

Figure 6.3: Position of streamers in relation to strip width on the ground when the aircraft is parked (source: Norton-Griffiths, 1978).

3. Place a second marker on the ground (B) so that w = W x h/H,

   whereby:

   w is transect width on the ground in meters,

   W is transect width required during the survey in meters,

   h is the height from the observer's eye to the ground in meters, and,

   H is the flying height during the survey in meters.

4. Position the second streamer on the wing strut (o) in the line of sight to point B (Figure 6.3).

5. Use a piece of tape to make a second mark on the window.

The formula above can be easily understood in terms of ratios. When measurements and calculations have been done correctly, the ratio of the transect width, when the aircraft is parked, to the transect width in flight equals the ratio of the height from the observer's eye to the ground, when the aircraft is parked, to the height above ground in flight (w/W=h/H). However, small errors in these measurements will result in large errors in actual strip width. Therefore, prior to a survey, the transect width will have to be calibrated. A line of markers will have to be laid out on the airstrip, preferably using bright objects placed 10 metres apart. The line of objects should be at least 50% longer than the required transect width. While using a pressure altimeter to determine height above ground level, the pilot should make a low pass over the markers to set the altimeter at zero. Then, using a radar altimeter, the pilot should make at least 10 passes over the markers at the required flying height. During each pass, the observers count the number of markers between the streamers. If it appears that the streamers are not positioned correctly and the difference between the required and the actual transect width is too great, than the pilot should land to reposition the streamers. If the error margin is acceptable, the actual transect width can be calculated with the above formula, whereby *h* is the average height of the aircraft during *x* passes, *w* is the average strip width during *x* passes, *H* is the required flying height during the survey and *W* is the actual strip width at that height (Norton-Griffiths, 1978). During flight, it is important that the observers keep the window tape and streamer lined up.

6.4 Analysis

Analysis methods for estimating population abundance from ground counts or aerial counts were first documented in the late 1960s (Jolly, 1969). The same methods of analysis are still used; e.g. Jolly's method 1 for equal-sized sample units (for example transects with the same length), and Jolly's method 2 for unequal-sized sample units (transects vary in length), sometimes referred to as the ratio method (Jolly, 1969).

Manual analysis of aerial survey data is simple, but time-consuming. If a computer is available, this process can be simplified and speeded up by using the program in conjunction with the brief manual provided in

Appendix I (Jachmann, 1995b). The program is written in Quickbasic 4.5, and will run under DOS on any computer.

6.4.1 Jolly's Method 1 for Equal Sized Sample Units

Jolly's method 1 is used when sample units have the same size; e.g. transects with the same length, or blocks that are all of the same size. The calculations are simple, because we do not need to know the size of the survey area and the population variance is the variance in animals counted between sample units multiplied by a simple weighting factor. The formulae for calculating the population estimate and precision are shown in Box 2.2. Working up the results of a survey with equal-sized sample units requires the following calculations:

1. Measure the length of the base-line from the map,

2. Determine the actual width of the transect or the size of the block,

3. Calculate the total number of sample units in the population (N), by dividing the base-line by the transect width, or by dividing the size of survey area by the size of the blocks,

4. Add the observations from the left and right observers for each sample unit (y_i),

5. Enumerate the observations of all sample units (Σy_i),

6. Calculate the sample mean: $\hat{y} = \Sigma y_i/n$,

7. Calculate the sample variance:

$$s_y^2 = \frac{1}{n-1} \times \{\Sigma y^2 - \frac{(\Sigma y)^2}{n}\}$$

8. Calculate the population total: $Y = N \times \hat{y}$

9. Calculate the population variance:

$$\mathrm{Var}(Y) = \frac{N(N-n)}{n} \times s_y^2$$

10. Calculate the population standard error: $SE(Y) = \sqrt{Var(Y)}$,

11. Calculate the 95% Confidence Limits: 95% CL = $\pm\, t \times SE(Y)$

6.4.2 Jolly's Method 2 for Unequal Sized Sample Units

The calculations for Jolly's method 2 are slightly more complicated. Because the sample units differ in size (transect lengths vary, or blocks vary in size), the population variance is the co-variance of the variance of animals counted in the transects or blocks and the variance between areas of the transects (length x width) or blocks. The formulae for calculating the population estimate and precision are shown in Box 6.1. Procedures and calculations will be discussed in section 6.6, using an example from the field.

Box 6.1: Formulae to calculate population estimate and precision, using Jolly's method 2.

N is the number of sample units in the population,
n is the number of sample units counted,
Z is the size of the survey area (km²),
z_i is the size of the ith sample unit,
y_i is the number of animals counted in the ith unit,
R is the ratio of animals counted to area covered $(\Sigma y_i / \Sigma z_i)$.

The population estimate: Y = Z x R

The variance between animals counted:

$$s_y^2 = \frac{1}{n-1} \times \left\{ \Sigma y^2 - \frac{(\Sigma y)^2}{n} \right\}$$

The variance between areas of sample units counted:

$$s_z^2 = \frac{1}{n-1} \times \{\Sigma z^2 - \frac{(\Sigma z)^2}{n}\}$$

The co-variance:

$$s_{zy} = \frac{1}{n-1} \times \{\Sigma(z \times y) - \frac{(\Sigma z) \times (\Sigma y)}{n}\}$$

The population variance:

$$Var(Y) = \frac{N(N-n)}{n} \times (s_y^2 - 2 \times R \times s_{zy} + R^2 \times s_z^2)$$

The population standard error: $SE(Y) = \sqrt{Var(Y)}$

The 95% CL: $\pm t \times SE(Y)$

Prior to the introduction of the GPS, the main problem with Jolly's method 2 was estimating transect length, and stratum size. Generally, method 2 is preferred, because it tends to produce more precise estimates.

6.5 Precision in Relation to Density and Sampling Intensity

The ideal of any survey is to obtain an estimate that is both accurate and precise. The accuracy of an estimate depends upon the species, the vegetation density and topography of the survey area, the survey design and operational factors. These variables act together to determine the size of the population that is visible to the observers at the time of the survey. At the lower end of the scale only a fraction of the population may be visible to the observers, whereas at the opposite end of the scale most of the population may be visible. When a large, conspicuous and gregarious species such as elephant is counted in wide open country, a high proportion of the population is visible. When bias is minimised by keeping operational factors such as speed and height of the aircraft within acceptable limits, and the strip width does not exceed 200 m, the resulting estimate may be accurate. However, when a small solitary species such as the common duiker is counted in a woodland area, the visible population is small and aerial counts will return underestimates. As a rule of thumb, in woodland savanna or other types of vegetation that obstruct visibility, one should never use aerial techniques to count solitary or small species.

The precision of a survey estimate is a measure of confidence in the result. It depends on the density of the target species, its spatial distribution, the survey design and the sample size (sampling intensity or % coverage of the survey area). With prior information on approximate density and distribution, the survey design and the sample size can be chosen to optimise the precision of the estimate. In a single species count, uneven distribution warrants the use of a stratified design to improve precision. With an even distribution, however, manipulation of the sample size is the only tool available to optimise the precision of the estimate. The survey results for elephant collected in the central Luangwa Valley between 1993 and 1999 will be used to illustrate the relationships between density, sampling intensity and precision.

6.5.1 Density and Precision

Generally, the spatial distribution and mean group size of elephants vary seasonally, depending on water and food availability. Human disturbance, such as poaching is important too. Elephants are gregarious, obtaining many social benefits from the formation of large herds (Moss and Poole, 1983; Dublin, 1986). During the wet season, the ample availability of preferred forage therefore results in large aggregations, while food competition during the dry season may limit herd size. Disturbance from human activity will cause compression of the population, due to collapse of the social structure (Lewis, 1986; Jachmann, 1988b), thereby locally increasing densities and group size. Compression can also be caused by elephants moving from a large disturbed area into a smaller quiet area, such as a well-protected National Park. This leads to an overall density increase in the quiet area. Increased elephant densities will result in larger groups, and the increment in group size will be determined by the level of harassment. Under these circumstances, elephants will be unevenly distributed and survey results will be less precise.

Density estimates and mean group sizes were collected during a series of 7 aerial surveys, carried out in the central Luangwa Valley from 1993 to 1999. Five surveys were flown in October and 2 in July (dry season). The surveys covered 2 National Parks (SLNP and Luambe) and parts of 6

GMAs (West Petauke, Chisomo, Sandwe, Lupande, Lumimba and Munyamadzi) adjacent to SLNP. In 1994, an additional 3 areas were covered; e.g. Lukusuzi and North Luangwa NPs and Musalangu GMA. The different conservation areas formed the strata. Breeding herds and male groups were combined when calculating the mean group size.

The level of elephant poaching in a particular area depends on the effectiveness of law enforcement, which is in turn determined by funding (Jachmann, 1998). In SLNP, between 1993 and 1999, between US$30 and US$40/km² per annum was spent on law enforcement, whereas adjacent conservation areas only received between US$2 and US$4 per km² per annum (Jachmann, 1998). During this period, elephant poaching was low in SLNP, with less than 0.25% of the population killed per year, but continued at higher rates in adjacent areas (Jachmann, 1998 and unpublished data). As a result, in SLNP the distribution of elephant was not affected by poaching and remained more or less even throughout the study period. Densities increased from 0.5/km² in 1993 to roughly 0.9/km² in 1996, declining to approximately 0.7/km² in 1999 (Jachmann and Phiri, 1999b). During this period, the mean group size increased with increasing elephant density (Figure 6.4). At low densities of up to approximately 0.65/km² there appeared to be little to no change in the mean group size of about 4. However, at densities higher than 0.65/km², the mean group size increased from about 4 to between 5 and 6 (Figure 6.4). In conservation areas adjacent to SLNP, the distribution of elephants was more clumped, with a higher mean group size starting at about 6 at densities of up to 0.4/km², increasing to approximately 14 at higher densities (Figure 6.4).

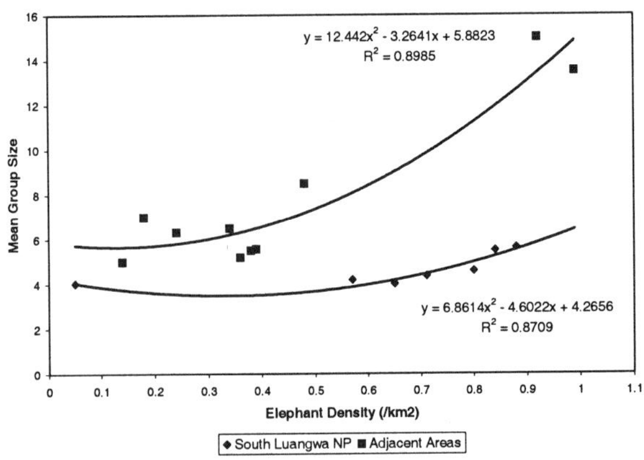

Figure 6.4: Relationship between mean group size and elephant density in South Luangwa National Park and adjacent areas, obtained through a series of 7 aerial surveys, carried out between 1993 and 1999.

Although these particular relationships are unique for the elephants in the central Luangwa Valley, the forming of larger groups with increasing densities has been documented before. In the early 1960s, the elephant population in the northern Serengeti (Tanzania) had a mean group size of 52.30 at 0.263/km², whereas the population in the south had a mean group size of 19.26 at 0.104/km² (Watson and Bell, 1969). During the same period, in North Bunyoro (Uganda), the elephant population in the high-density stratum (3.8/km²) had a mean group size of 22.5, whereas the population in the medium-density stratum (1.9/km²) had a mean group size of 6.6 (Laws *et al.*, 1975).

For an elephant survey done at a particular sampling intensity, the precision of the estimate will decline with increasing numbers, hence with larger group sizes. This relationship is shown in Figure 6.5, where estimates for areas sampled at low intensities (<4%) and those with high poaching levels were omitted.

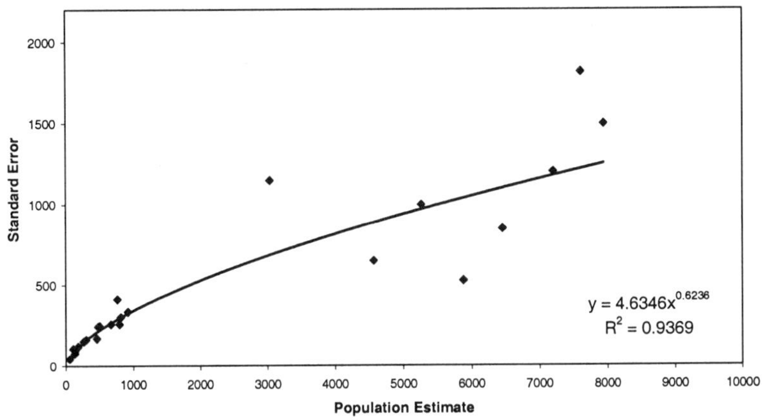

Figure 6.5: Relationship between precision and numbers.

The important question is why precision declines with increasing group size? The answer can be deduced from the relationship between density and mean group size (Figure 6.4). The mean group size increases above densities of 0.65/km² in areas in the central Luangwa Valley with low levels of poaching and above densities of 0.4/km² in areas with high levels of poaching. In other words the size of groups increases at a faster pace than density increases, whereby the space between different groups also increases. The net effect is fewer groups with more elephants per group, which means the population is more clumped. This effect is illustrated in Figure 6.6, where 6 transects are placed in a hypothetical survey area; first in a low-density situation with 17 small groups of elephants (A) and second in a high-density situation with 10 large groups of elephants (B).

Without taking account of differences in group size, situation A will result in 1.50 ± 0.55 groups per sample unit (SE is 36.67%) and situation B in 0.83 ± 0.75 groups (SE is 90.36%). Thus, surveys in B will be less precise than in A.

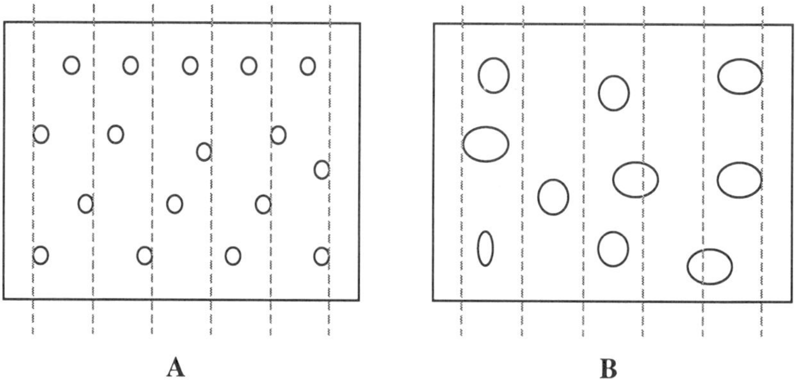

Figure 6.6: Illustration of the effect of mean group size on the precision of a survey result.

6.5.2 Sample Size and Precision

The precision of a survey result is not only related to the density or mean group size, but also to sampling intensity. Because poaching results in compression of the population and consequently in locally higher densities and group sizes, only survey results for the SLAMU area were used to illustrate this point (SLNP and Lupande GMA). The relationship follows a power curve, whereby precision improves rapidly with increasing sampling intensity (Figure 6.7). The data show that we require at least 5% coverage to obtain an adequate level of precision (Figure 6.7). Note that the law of diminishing returns operates here: sampling intensities of 10% require more effort but bring little improvement in precision (Figure 6.7).

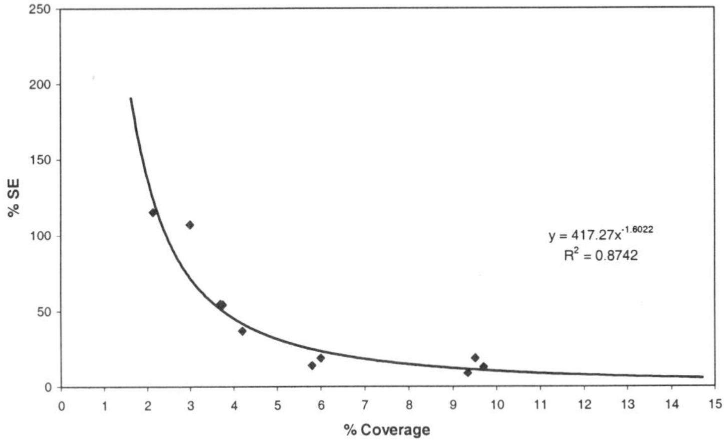

Figure 6.7: Relationship between precision and sampling intensity for a series of elephant surveys carried out in SLNP and Lupande GMA, from 1993 to 1999.

6.6 Field Example of an Aerial Survey

An example from the field will be used to illustrate the procedures and calculations for working up the results of an aerial sample count. The data will be analysed using Jolly's method 2 for unequal-sized sample units (Jolly, 1969). For convenience, notations will follow Norton-Griffiths (1978).

Table 6.1: Example of data sheet used for aerial surveys in the Luangwa Valley.

AERIAL SURVEY DATA SHEET	**Ref:.......**
Date	:..
Area/Stratum	:..
Species	:..
Pilot	:..
Navigator	:..
Observer:	
Mid-Left	:..
Mid-Right	:..
Back-Left	:..
Back-Right	:..
Transect #:..................	
Time Start:..................	

Time	Species	#	# Film	# Frame	Dead Elephant (O or F)*
....	....	....	....	....	....
....	....	....	....	....	....
....	....	....	....	....	....
....	....	....	....	....	....
....	....	....	....	....	....
Time Finish:.......................					
***O is bones only, F is fresh with skin still visible**					

These data were collected during a survey of SLNP in 1999. Forty-three transects were flown between the Muchinga Escarpment and the Luangwa River (*see Figure 5.4*). Nominal flying height was 450 feet a.g.l and the nominal transect width was 200 m on both sides of the aircraft. Transect width was calibrated according to the procedure described in section 6.3. A GPS was used for navigation and a radar altimeter for height control. The double-count procedure was followed, with two additional observers in the back of the aircraft. Sightings were recorded on data sheets, including the exact time for each observation (Table 6.1). Also, dead elephants were recorded, divided into old carcasses (only bones visible) and fresh carcasses (skin visible). Large herds were supposed to be photographed. Prior to each flight, the watches of all observers were synchronised to facilitate the double-count procedure and to obtain accurate co-ordinates for each sighting. The co-ordinates of sightings were later plotted on a map to determine dry-season distribution.

These steps should be followed, using an ordinary calculator:

1. When the survey included a double count to correct for visibility bias, calculate a correction factor for each species counted (C_f), using the observations from one or two additional observers. Procedures and formula for the double-count technique are provided in section 5.3.

The correction factor for elephants for the 1999 aerial survey was 1.15, implying that 15% of the visible population was missed by the two front observers.

2. Calculate the size of the survey area or stratum (Z). This is found directly from the map by superimposing a grid and counting the

grid squares, or by simply calculating the size of the area using measurements from the map.

In our example, the size of the survey area or stratum is the whole SLNP, which is 9,050 km².

3. Measure the length of the base-line from the map.

The base-line for SLNP was 106 km.

4. Prepare a data sheet along the lines of Table 6.1.

5. Calculate the actual transect width (W).

In our example, the calibrated transect width (see section 7.3 for calibration procedure) was 200.0 m for the left strip and 181.1 m for the right strip, giving a nominal transect width of 381.1 m. The nominal flying height was 450 feet, which is 450 x 0.3048 m/foot = 137.2 m. The actual transect width during the survey depends on the actual flying height, which is found by averaging all the recordings from the radar altimeter (see section 6.3). During the survey of SLNP, average height was 451.7 feet, which is 137.7 m. Then, the actual transect width is (137.7/137.2) x 381.1 = 382.5 m.

6. Calculate the total number of sample units available, from which the transects to be flown will be selected (N). This is done by dividing the length of the base-line by the actual strip or transect width.

For SLNP; $N = 106/0.3825 = 277.12$.

7. Determine the length (l_i) of each transect (n_i) from measurements from the map (*Table 6.2*).

8. Calculate the area of each transect (z_i) by multiplying the length with the actual transect width; $z_i = l_i \times W$ (*Table 6.2*). Sum the areas to give the total area sampled ($\sum z_i$).

In our example, the total area sampled is 849 km² (Table 6.2).

9. Correct for counting bias by using the photographs of large herds.

In our example, elephant groups were never large enough to necessitate the use of photographs.

10. Add the data from both front observers (y_i is observations left strip + observations right strip), and enumerate animals counted in the sample (Σy_i).

($\Sigma y_i = 463$; Table 6.2).

11. Calculate R, which is the ratio of animals counted to area covered. R provides an estimate of the density of animals in the sample area ($R = \Sigma y_i / \Sigma z_i$).

In our example $R = 463/849 = 0.545$ elephants/km².

12. Calculate the population total (Y) by multiplying the density ® by the observer-bias correction factor (C_f) and the size of the survey area or stratum (Z); $Y = R \times C_f \times Z$. In a survey without a double count C_f is 1.

In our example $Y = 0.545 \times 1.15 \times 9{,}050 = 5{,}672$ elephants in SLNP.

13. Calculate the variance between animals counted in the transects: $s_y^2 = (1/n\text{-}1) \times (\Sigma y^2 \times C_f^2 - (\Sigma y_i \times C_f)^2/n)$.

In our example $s_y^2 = 1/42 \times (10{,}773 \times 1.32 - (463 \times 1.15)^2 /43) = 181.60$.

14. Calculate the variance between areas of the transects: $s_z^2 = (1/n\text{-}1) \times ((\Sigma z^2) - (\Sigma z)^2/n)$.

In our example $s_z^2 = 1/42 \times (18{,}065 - 720{,}801/43) = 31.00$.

15. Calculate the co-variance between the animals counted and the areas of the transects: $s_{zy} = (1/n\text{-}1) \times (\Sigma z.y \times C_f - (\Sigma z \times \Sigma y \times C_f)/n)$.

In our example $s_{zy} = 1/42 \times (11{,}493 - 452{,}050/43) = 23.34$.

16. Calculate the population variance: $Var(Y) = (N(N\text{-}n)/n) \times (s_y^2 - 2 \times R \times s_{zy} + R^2 \times s_z^2)$.

In our example *Var(Y) = ((277.12(277.12 – 43))/43)* x *(181.60 – 2* x *0.545* x *23.34 + 0.297* x *31.00) = 1,508.82* x *(181.60 – 25.44 + 9.21) = 249,513.56.*

17. Calculate the population standard error: *SE (Y) =* √ *Var(Y).*

In our example SE(Y) = 500.

18. Calculate the 95% confidence limits of Y: CL = ± *t x SE(Y),* where t depends on the number of transects in the unit (n). When n > 30, t is 1.96.

In our example, the 95% confidence limits are 1.96 x 500 = ± 980.

Thus, the number of elephants in SLNP is 5,672 ± 500 (%SE = 8.82%); (95% CL 4,692 – 6,652). This implies that SLNP definitely has at least 4,692 elephants, and possibly as many as 6,652, with the real population size probably around 5,672 (see section 2.6)

When using blocks instead of transects, the calculations are essentially the same. The size of the sample units (z_i) is the size of the blocks and the total number of sample units in the population (N) is the survey area divided by the mean block size ($N = Z/(\sum z_i / n)$).

Table 6.2: Length, width and area of 43 transects placed in SLNP, the number of elephants counted in each transect and the calculations required to estimate the variance, standard error and confidence limits.

Tr. n_i	**Length (km)** l_i	**Width (km)** W	**Area (km²)** z_i	y_i	z_i^2	y_i^2	z_i x y_i
1	12.8	0.3825	4.896	0	23.971	0	0
2	19.7	0.3825	7.535	13	56.776	169	97.955
3	27.2	0.3825	10.404	38	108.243	1444	395.352
4	31.3	0.3825	11.972	13	143.329	169	155.636
5	36.0	0.3825	13.770	0	189.613	0	0
6	41.4	0.3825	15.836	28	250.779	784	443.408
7	45.3	0.3825	17.327	0	300.225	0	0
8	51.3	0.3825	19.622	0	386.594	0	0
9	58.5	0.3825	22.376	5	500.685	25	111.880
10	64.7	0.3825	24.748	18	612.464	324	445.464
11	68.8	0.3825	26.316	1	692.532	1	26.316
12	68.8	0.3825	26.316	21	692.532	441	552.636
13	66.2	0.3825	25.322	8	641.204	64	202.576
14	66.7	0.3825	25.513	16	650.913	256	408.208
15	62.1	0.3825	23.753	20	564.205	400	475.060

16	61.0	0.3825	23.332	36	544.382	1296	839.952
17	61.5	0.3825	23.524	12	553.379	144	282.288
18	64.5	0.3825	24.671	35	608.658	1225	863.485
19	64.3	0.3825	24.595	41	604.914	1681	1008.395
20	76.0	0.3825	29.070	11	845.065	121	319.770
21	81.9	0.3825	31.327	32	981.381	1024	1002.464
22	63.0	0.3825	24.098	18	580.714	324	433.764
23	61.1	0.3825	23.371	16	546.204	256	373.936
24	56.7	0.3825	21.688	10	470.369	100	216.880
25	52.4	0.3825	20.043	2	401.722	4	40.086
26	53.3	0.3825	20.387	6	415.630	36	122.322
27	52.8	0.3825	20.196	4	407.878	16	80.784
28	50.4	0.3825	19.278	3	371.641	9	57.834
29	52.8	0.3825	20.196	0	407.878	0	0
30	52.4	0.3825	19.278	5	371.641	25	96.390
31	52.2	0.3825	19.966	4	398.641	16	79.864
32	51.5	0.3825	19.699	7	388.051	49	137.893
33	46.3	0.3825	17.710	6	313.644	36	106.260
34	45.0	0.3825	17.212	12	296.253	144	206.544
35	48.0	0.3825	18.360	10	337.090	100	183.600
36	49.6	0.3825	18.972	3	359.937	9	56.916
37	51.5	0.3825	19.699	0	388.051	0	0
38	52.0	0.3825	19.890	9	395.612	81	170.010
39	52.0	0.3825	19.890	0	395.612	0	0
40	51.0	0.3825	19.508	0	380.562	0	0
41	41.0	0.3825	15.682	0	245.925	0	0
42	32.7	0.3825	12.508	0	156.450	0	0
43	23.9	0.3825	9.142	0	83.576	0	0
n = 43			Σz_i = 849	Σy_i= 463	Σz_i^2 = 18065	Σy_i^2 = 10773	$\Sigma z_i \times y_i$ = 9994

6.6.1 Analysis of a Stratified Design

Analysis of a stratified design requires the population estimates, the population variances and the sample sizes of each stratum. As an example, the survey results collected in the central Luangwa Valley in 1999 will be used (Table 6.3). The valley was stratified by conservation area. Sampling intensity for each stratum was that used in previous years.

The total estimate is obtained by summing the estimates for the individual strata, while the precision is estimated by summing the variances and calculating the square root of the sum.

The sum of the individual stratum estimates was 9,162 (Table 6.3). The total variance was 924,332, so the SE = √924,332 = 961. Therefore the 95% CL were 1.96 x 961 = 1,884.

Thus, in the central Luangwa Valley there are 9,162 ± 961 (%SE = 10.49%) elephants (95% CL 7,278 – 11,046). In the central Luangwa Valley there are definitely at least 7,278 elephants, and possibly as many as 11,046, with the real population size probably around 9,162 (see section 2.6).

Table 6.3: Aerial survey results for the central Luangwa Valley, collected in 1999.

Stratum	Size (km^2)	Estimate	Variance	Sample
South Luangwa NP	9,050	5,672	249514	43
Lupande GMA	1,489	493	33801	13
Munyamadzi GMA	2,200	919	45393	16
Lumimba GMA	2,415	1,053	79004	16
Sandwe/Chisomo GMAs	750	128	6195	7
West Petauke GMA	905	897	510425	4
Total	16,809	9,162	924332	99

6.7 Comparative Costs of Aerial Counts

The cost of an aerial sample count is mainly determined by the cost of running or hiring an aircraft and the sampling intensity of the survey. Aerial counts in Zambia are mainly done with hired aircraft, at a rate that is more or less representative for most of the continent (US$ 220/hour, which includes fuel). The survey of the central Luangwa Valley carried out in 1999 covered a total area of 16,809 km^2 with an average sampling intensity of about 10% and a total cost of US$ 13,120 or US$ 0.78/km^2 (Table 6.4). The survey took approximately 10 days to complete. Approximately 80% of the expenditure was aircraft hire (47 hours), which includes placement time. The remaining 20% was vehicle running and staff allowances. If one were to conduct an aerial total count of the same area, only counting large conspicuous ungulates such as elephants, buffalo and wildebeest, the maximum allowable transect width would be 1 km (500 m + 500 m). This implies a coverage of factor 2.5 times that of the sample count, with a total cost of US$ 32,800, or US$ 1.95/km^2 (Table 5.4). The time involved to complete the census with a single aircraft would be approximately 25 days. A line-transect ground survey of the central Luangwa Valley would cost US$ 75,640, or US$ 4.50/km^2 (Table 6.4). Using 6 teams of 3 people each, a line transect survey would take roughly 3 months to complete to obtain results with an adequate level of precision.

Thus, aerial sample counts are time and cost-efficient compared with total counts and line-transect ground counts. A major disadvantage of aerial sample counts is that in woodland areas they return poor information on smaller species compared with line-transect ground counts. But that has to be balanced by the fact that line-transect counts are 6 times more expensive and require 10 times more time than aerial sample counts.

Table 6.4: Cost of aerial sample count done in the central Luangwa Valley in 1999 (16,809 km^2), compared with the cost of an aerial total count and a line-transect ground count covering the same area.

	Aerial Sample	**Aerial Total**	**Line Transect**
Total Cost	US$ 13,120	US$ 32,800	US$ 75,640
Cost per km^2	US$ 0.78	US$ 1.95	US$ 4.50

PART III

INDIRECT COUNTS

PART IIIa

SPECIAL AND INDEX TECHNIQUES

Chapter 7

Special Techniques

Certain indirect techniques, such as the mark/recapture method, or estimating abundance through radio tracking or by using DNA analysis, are special cases. Although these methods are used when only part of a population is visible to the observers, they do not come under the category of index techniques, because they lead to estimates of population size.

7.1 The Mark/Recapture Method

There exists ample literature on this technique, summarised by Seber (1982, 1986, 1992) and Lebreton *et al.* (1992). Although the technique can be used to examine a variety of properties of an animal population, the discussion will be restricted to a brief introduction on estimation of population size.

In the mark/recapture technique, a sample of animals is captured, marked and released. Then, using sample theory and the properties of this marked sample, the size of the population can be estimated from a sample of recaptured animals. Sample theory states that the ratio of the number of marked animals (M) to the total number of animals in a closed population of unknown size (N) equals the ratio of m marked individuals in a subsequent sample of n animals, or:

$$\frac{M}{N} = \frac{m}{n} \quad or \quad N = \frac{Mn}{m}$$

Although the general theory and application of the method are relatively simple, the experimental design is critical. First, it is crucial to the reliability of the results that all individuals in the population, both marked and unmarked, are equally catchable. Second, the population must occupy a distinct area with known boundaries, and third, no marks should be lost. Last but not least, during the experiment the population must be stable (e.g. no migration, no recruitment and no mortality). Physically marking a sufficient number of large African ungulates in a particular population is a tedious and expensive task. However, an indirect method of marking wild animals exist, such as individual recognition of a number of animals in a population. Specific characteristics of tusks and ears in elephants, colour patterns in giraffe and colour patterns and horn shape in nyala antelope are some of the examples that have been used for individual recognition in the past. A major disadvantage of mark/recapture through individual recognition is the time required. However, the advantage is that marks cannot be lost. In any case, even when a satisfactory experiment can be designed, estimating numbers of large African ungulates by mark/recapture will always be labour intensive and therefore expensive. Also, results may be inaccurate. Particular conditions may exist, however, where mark/recapture is the only option available, because the use of simpler and more reliable methods is not feasible.

7.2 Field Example

One of the few examples documented for Africa is the estimation of abundance of nyala antelope (*Tragelaphus angasi*) in the small Lengwe National Park in the Lower Shire Valley in Malawi. In the early 1980s, roughly 3,000 to 4,000 nyala occupied an area of approximately 200 km² of predominantly thicket vegetation (Bell, 1981; Bell and Banda, 1983). Serious degradation of the thicket community necessitated a series of culling operations, starting in 1981 (Mphande and Jamusana, 1984). A prerequisite to any culling operation is the availability of sound information on the abundance of the target species. Due to the density of the thicket vegetation, nyala could not be counted with any of the conventional techniques, leaving the authorities with only one valid option, the use of the mark/recapture method. During the peak dry season, only four waterholes remained, allowing observers to indirectly mark nyalas by individual recognition, by observations from a hide near one of the four waterholes. The exercise was confined to the 100 km² of 'old Lengwe National Park', and took 2 weeks to complete. The first week was used to build up a recognition file, using a binocular telescope. Nyala are easily recognised because of the spot/stripe pattern. However, with 20 – 30 nyala drinking at once, obvious distinguishing marks were used, such as ear notches, broken horns, scars or stripe patterns. A sketch was made of a selected sample of the most easily recognised animals. The second week was devoted to recognising and counting nyala coming to the waterholes. When the sample of animals "marked" through individual recognition was sufficiently large, the size of the population was estimated using the Petersen Estimate (Seber, 1982; Caughley, 1977).

The Petersen Estimate is the most frequently used equation in mark/recapture estimates of abundance, mainly because it is simple and relatively reliable. Using the method of individual recognition, where in principle no marks are lost, the Petersen Estimate yields accurate results when the following conditions are met:

1. The probability of indirectly marking an animal through individual recognition is the same for all individuals in the population. *In Lengwe National Park this condition existed, because each animal had to visit one of the four water holes at regular intervals, and all water holes could be monitored.*

2. There should be no recruitment through births and immigration between marking and recapturing. *Lengwe National Park forms a small island in a sea of cultivation, and movement of nyala, either away or into the study area, was unlikely. The condition of recruitment through births was not violated, because sampling was during the peak dry season, between birth pulses.*

3. Mortality of individually recognised animals is the same as that of other animals in the population. *In Lengwe National Park, there was no reason to assume that mortality rates differed between "marked" and "unmarked" individuals.*

Using the mark/recapture method with individual recognition, the number of "marked" individuals is exactly known prior to recapturing. This enables a decision on the number of animals to be recaptured for a particular level of precision. The size of the population can be estimated with the formulae provided by Seber (1982), shown in Box 7.1:

Box 7.1: Formulae for calculating population estimate and precision, using the mark/recapture method (Seber, 1982).

The population estimate is given by:

$$Y = \frac{n(M + 1)}{m - 1}$$

where:

Y is population estimate,
M is number of "marked" individuals,
n is total number of individuals recaptured,

m is number of "marked" individuals recaptured.

The population variance by:

$$Var(Y) = \frac{(M - m + 1)(N + 1)(Y - M)}{m(m + 2)}$$

The population standard error: $SE(Y) = \sqrt{Var(Y)}$

The 95% CL: ± t x SE(Y)

The number of animals that are recognised individually and that must be recaptured to obtain a particular level of precision can be approximated from an educated guess of the size of the population using the following equation (Caughley, 1977):

$$m = \frac{SQ\{\{\frac{(Y+1)(Y-M)}{SE^2}+2\}^2+\{\frac{4(Y+1)(Y-M)(M+1)}{SE^2}\}-\{\frac{(Y+1)(Y-M)}{SE^2}+2\}\}}{2}$$

where:

SQ is square root

SE is standard error

To illustrate the use of the formula above, let us assume that the management of Lengwe National Park suspected a population of roughly 3,000 nyala, based on patrol reports and general observations at waterholes. Exactly 100 nyala could be recognised individually and the culling exercise required a population estimate with a standard error of 5% or roughly 5 x 30 = 150. Using the above equation (Caughley, 1977), the number of nyala that must be recaptured is estimated at 277, which is 177 more than can be individually recognised. If the required precision was 10%, which is 10 x 30 = 300, the number of nyala that must be recaptured is 110, still 10 more than can be recognised. With a precision of 15%, which is 15 x 30 = 450, the number of nyala that must be recaptured is 70, 30 less than can be recognised. Thus, the choice is between increasing the number of "marked" nyala, or accepting a population estimate with a standard error of more than 5%. Suppose that a lower level of precision is accepted and only 70 "marked" individuals are recaptured (m) of a total sample of 1,600 nyala (n), then the population size is estimated as (see Box 7.1):

$$Y = \frac{1,600(100 + 1)}{70 - 1} = 2,342$$

with a standard error of:

$$SE(Y)=SQ\{\{\frac{(100 - 70 + 1)(2,342 + 1)(2,342 - 100)}{70(70 + 2)}\}\}=180=7.6\%$$

The precision of the estimate is better than expected, mainly because the population estimate was considerably lower than 3,000. Simplifying the experiment we can state that if we recapture 22.86 times more "unmarked" than "marked" nyala (1,600/70 = 22.86) in a closed population with exactly 100 "marked" nyala, the true number of nyala must be approximately 22.86 x 100 = 2,286. As a general but rough guideline, in small populations (< 3,000) we should recapture approximately half of the "marked" individuals for an estimate with a standard error of 10% or less.

The most recent application of the mark/recapture technique has been the monitoring of black rhino in the Hluhluwe-Umfolozi National Park in South Africa (Galli and Flamand, 1995). Here, individual recognition and ear notches were used to mark a proportion of the population. The population size was estimated with Bayesian mark/recapture analysis, using the 'RHINO' computer programme (Emslie, 1993).

7.3 Estimating Abundance through Radio-Tracking

Estimating abundance through radio-tracking is also a special case. This technique is mainly used to study movements of long-ranging animals, such as elephants. However, it can also be used to estimate the average activity area or home range of a sedentary species that is difficult to count with other methods. An estimate of abundance can be obtained once the activity areas have been estimated for a number of individuals or groups. Assuming there is little to no overlay between activity areas and the species is evenly distributed, then the density in the activity areas can be extrapolated to the whole area. Estimates of economically important predator species, such as lion and leopard, can be obtained through individual recognition, some type of index count, or by radio telemetry. Radio telemetry involves putting a small VHF radio transmitter on a number of selected animals. The transmitter is attached to a collar, and emits a pulsing signal on a predetermined frequency. This signal can be detected on a VHF receiver set to the same frequency. The transmitter, and therefore the study animals, can be located with a directional antenna. When the antenna is directed straight at the transmitter, the signal becomes stronger. The signal strength also increases as the observer approaches the animal. The main disadvantage of this method is that it is labour-intensive and expensive.

There are essentially two approaches to determining the position of a radio-collared animal:

- In the first approach, the position of the animal is determined by triangulation. With triangulation, the observer first determines his own position on the map. Then, a compass bearing is taken of the source of the signal. On the map, a line is drawn from the observer's position to the source of the signal. The same process, but from a different location, will give two converging lines on the map (Figure 7.1). Where the two lines converge is the approximate position of the animal (*position fix*). Triangulation can be done by a mobile observer, or with several fixed base stations. This gives an estimate of the animal's position on a map, or a particular grid reference (*remote radio telemetry*).

- In the second approach, visual contact is made, allowing accurate plotting of the position, as well as collection of additional information on group dynamics, habitat and behaviour (*visual contact radio telemetry*).

Making visual contact and plotting the position with a GPS is definitely the preferred approach, because it eliminates errors associated with triangulation. If, however, this is not possible because the type of habitat limits the use of vehicles, or it is not advised to approach the study animal(s) on foot, remote radio telemetry has to be used.

The error in the method of visual contact is determined by the error in the GPS reading. The total error in the method of remote radio telemetry is to a large extent determined by system errors, errors caused by movement of animals between consecutive readings, and errors caused by habitat features (Macdonald and Amlaner, 1980).

System Errors: system errors are caused by damaged antennae, or by geometrical errors in triangulation. When taking a compass reading of the source of the signal, the true position of the animal may be several degrees either to the left or to the right. This error becomes larger with a greater distance between the observer and the animal (Figure 7.1).

Movement Errors: when the collared animal moves between two successive readings, the magnitude of the error is determined by the speed of the animal and the time lag between readings.

Habitat Features: VHF waves are attenuated by vegetation and reflected by hills and rocks. Therefore, these inaccuracies vary between areas.

Most of these errors can be minimised by using a non-linear least-squares estimator (Pace and Weeks, 1990). However, this procedure requires complicated trigonometry and matrix algebra, which is beyond the scope of this manual.

Whatever the method used, the data-set will consist of a number of fixes or co-ordinates (sample size), reflecting the different positions of the collared animals over the study area. Many different techniques have been devised to analyse home-range data, reviewed by Harris *et al.* (1990). Some of these techniques will be discussed, but first we need to look at the sample size required for analysis.

The sample size must be large enough to give an accurate estimate of the activity area. The optimum number of observations will vary according to the species and the local conditions. As the number of observations increases, so will the size of the activity area (Figure 7.2). Initially the increase in area size is rapid, until a point is reached where an increase in sample size does not result in an increase in estimated activity area (Figure 7.2). This asymptotic value is taken as the minimum sample size required, under those particular conditions.

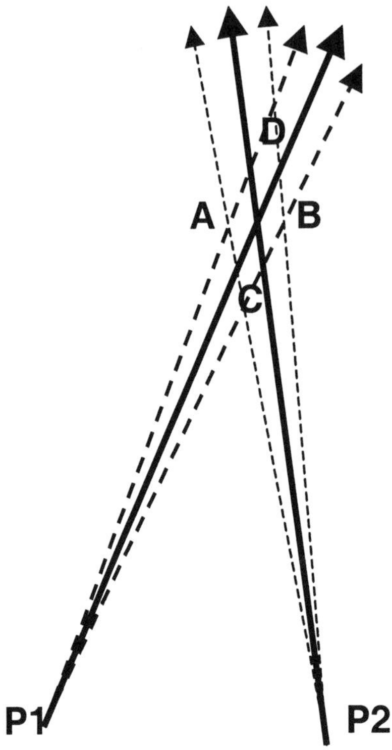

Figure 7.1: Geometrical errors in the method of triangulation. Compass bearings are taken at positions P1 and P2. The error is given by the area of the parallelogram ABCD.

There are many different ways of analysing and presenting activity area data. However, most statistical procedures are biased, but to a different extent (Boulanger and Whyte, 1990). More problems are created when data analysed with different statistical procedures and different sample sizes are compared (Boulanger and Whyte, 1990). Therefore, estimates of activity areas should be seen as a general measure only, and estimates of abundance derived from these as an approximation of true abundance. The simplest and most useful technique for estimating the size of an activity area is the *minimum convex polygon* (Figure 7.3). This is the only technique that is comparable between studies, because it does not rely on complex statistical procedures. The procedure is simply to map the co-ordinates and to draw straight lines between peripheral data points (Figure 7.3). The main disadvantage is that this method does not provide an indication of intensity of range use, because some areas within the polygon may seldom or never be visited (Whyte, 1996). Thus, the estimate of area size is often too high.

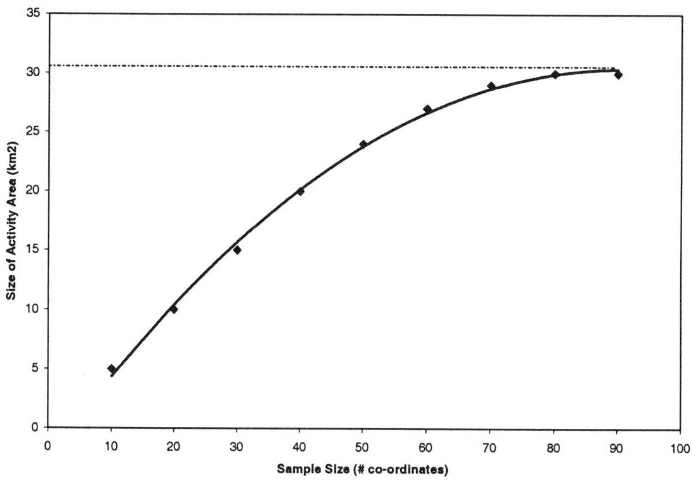

Figure 7.2: The size of the activity area increases with increasing sample size (number of observations). The point where a further increase in sample size does not give an increase in the size of the activity area, is known as the asymptote, which is the minimum sample size required.

Accuracy can be considerably improved by using a *sorted polygon* (Voigt and Tinline, 1980). After determining the theoretical centre point, which is the centre of the data-set when plotted on a map, each observation is numbered according to its direction from the centre of activity. The points are then joined in numerical order to give an irregular shaped area that can then be calculated.

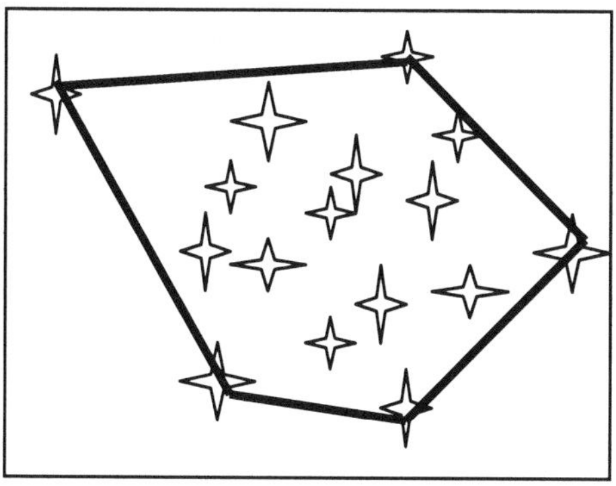

Figure 7.3: Simplified hypothetical example of a minimum convex polygon.

Another commonly-used non-statistical technique is to overlay grid cells, where area size is estimated by adding up the number of cells with one or more observations (Figure 7.4). When a single observation covers two or more adjacent cells, all cells involved are counted.

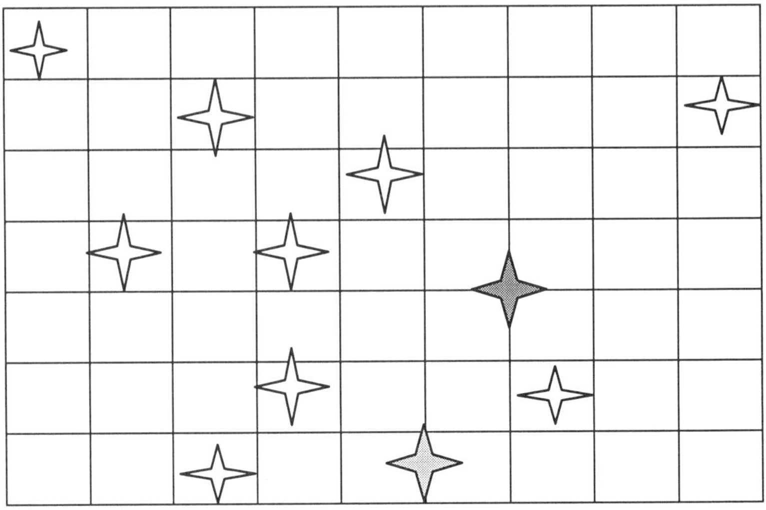

Figure 7.4: Simplified hypothetical example of the estimation of area size by overlaying grid cells. In this example, the activity area is estimated by 15 cells, 9 observations in single cells, one observation in two adjacent cells (light grey), and one observation in four adjacent cells (dark grey).

The more complicated statistical techniques either use activity areas assuming a bivariate normal distribution, or a uniform distribution (Figure 7.5).

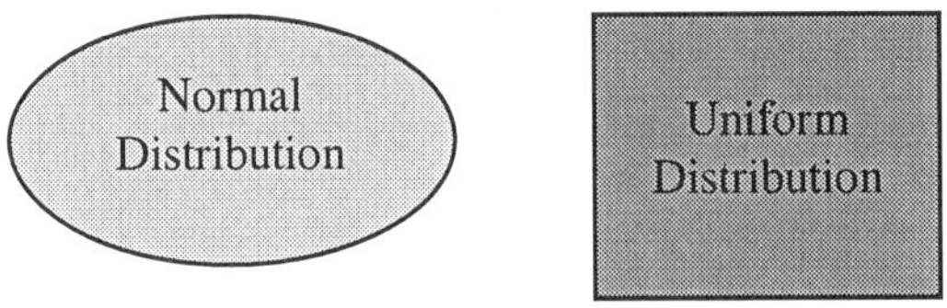

Figure 7.5: Activity areas with observations from a bivariate normal distribution (light grey) and from a uniform distribution (dark grey).

The following estimators are commonly used to calculate the size of an activity area: Koeppl *et al.*'s (1975) 95% ellipse, Jennrich and Turner (1969) 95% ellipse, Anderson's (1980) Fourier Series, and Dixon and Chapman's (1980) Harmonic Mean. All of these estimators can be computed with the McPAAL software package (M. Stüwe and C.E. Blohowiak, Conserv. Res. Cent., Natl. zool. Park, Smithsonian Inst., Front Royal, Va., 1985).

7.4 Field Example

To illustrate the use of this technique, we will discuss an example from the central Luangwa Valley, where between April 1990 and August 1991, lion density was estimated by visual contact radio telemetry (Yamazaki, 1992). The study area comprised the riverine part of the Nsefu Sector of South Luangwa National Park, and the riverine part of the Upper Lupande Game Management Area, totalling approximately 355 km^2 on the east bank of the Luangwa River. The area is characterised by large grassy plains with scattered *Capparis tomentosa* bushes, mopane (*Colophospermum mopane*) woodland, and *Acacia/Combretum* thickets.

In the Lupande GMA, the annual safari-hunting quota for adult male lions had been set at about 8 for many years. This quota was based on educated guesswork. Therefore, the objective of the study was to get a general idea of lion density in the riverine zone of the Lupande GMA, mainly for quota-setting purposes. In the Lupande GMA, safari hunting is mainly done in the alluvial valley trough, covering a block of about 1,660 km^2.

Between April and August 1990, 6 lions of 4 prides were tranquillised with a powder-charged dart-gun (Pneu-Dart Inc., #171c). The darts either contained a mixture of Ketamine and Xylazine or Tiletamine and Zolazepam. The lions were fitted with radio-collars from Telonics Inc. In the two large prides (Nos. 1 and 2), an adult female and an adult male were fitted with radio-collars, while in the two smaller prides (Nos 3 and 4), only a single adult female was equipped with a transmitter. During the tracking period, from August 1990 to August 1991, the 4 prides consisted of 22 ± 2 (1), 14 ± 1 (2), 5 ± 1 (3) and 3 ± 1 (4) lions on average. Sample sizes, or number of position fixes collected after visual contact, were 82 (1), 67 (2), 29 (3) and 36 (4) respectively (Table 7.1). Activity areas were estimated with minimum convex polygons (Table 7.1). There was some overlap among activity areas (Yamazaki, 1992).

During the study period, there were 44 ± 5 lions occupying an area of about 230 km^2 (Table 7.1). Overlap of activity areas was 52 ± 4 km^2, resulting in an effective range of 230 – 52 ± 4 = 178 ± 4 km^2. This gave an estimated lion density of 0.2472 ± 0.0287/km^2 (± SE). The riverine hunting block covers about 1,660 km^2. The number of lions in the hunting block was estimated at 1,660 x 0.2472 ± 0.0287 = 410 ± 48. During the study period, the ratio of adults to sub-adults was 1 : 1, and the ratio of adult males to adult females was 0.24 : 1 (Yamazaki, 1992). This results in 205 adults, with an estimated 40 ± 5 adult males.

Table 7.1: Radio-tracking results for 4 prides of lions in the central Luangwa Valley, showing the average number of animals in each pride, sample sizes, and the estimated size of the activity areas.

Pride	**Average # Individuals**	**Sample Size (# Observations)**	**Estimated Activity Area (km^2)**
1	22 ± 2	82	82.6
2	14 ± 1	67	53.9
3	5 ± 1	29	53.8
4	3 ± 1	36	39.6
Total	44 ± 5	214	229.9

Lions are capable of high population growth rates (10 - 15% per annum). They can sustain sport hunting quotas of up to 8% per annum. The annual quota for Lupande GMA had been 20% of the adult male population, which is much too high, and accounts for the low occurrence of good trophy quality lions in that area.

Figure 7.6: Japanese volunteer, Koji Yamazaki, searching for a radio-collared lion in the Lupande GMA.

7.5 Estimating Abundance using DNA Analysis

In this recently introduced technique, faecal or hair samples are analyzed for genetic markers. These genetic markers can be used to study taxonomy, population dynamics, food habits and population size. Although this technique uses indicators of animal presence (*see part IIIb*), it is a special case, because droppings and hairs are not counted but analyzed.

The technique allows for the purification of genetic material (DNA) from faeces or the root portion of hairs. This is followed by amplification of specific DNA sequences using the polymerase chain reaction (Kohn and Wayne, 1997). Genetic markers with different properties can then be used for individual identification, known as "genetic fingerprinting".

DNA profiling is a promising technique for estimating distribution and abundance of free-ranging populations, with a wide application in the field. The only disadvantage is that DNA analyses requires a sophisticated laboratory, and considerable financial input. Because of the cost involved, it is not expected that this method will be widely applied on the African continent.

The technique has amongst others been used to estimate the abundance of coyotes in the Santa Monica Mountains near Los Angeles, California, using faecal material (Kohn et al, 1999), and for grizzly bears in British Columbia and Alberta, Canada, using root portions of hairs collected at bait sites, combined with mark/recapture analysis (Mowat and Strobeck, 2000). DNA profiling of hairs can be used for studying large carnivores, and faecal material for most other species.

Chapter 8

Index Counts

Counting techniques discussed in previous chapters produce absolute estimates of density, whereas index counts yield relative densities or density indices. Relative densities can be defined as the density of one population relative to another, or the density of a population at time t relative to its density at time $t + 1$. Thus, relative densities are useful only in temporal and spatial comparisons, and only if the conditions and methodology used are exactly the same for the areas and time periods compared. When conditions and methodology are consistent, index counts are more cost-efficient than direct counts, and can be used for trend analysis or as feedback for law-enforcement operations. A density index should be a measurable correlative of density, preferably with a linear trend on absolute density. A linear relationship can be expressed as *Absolute Density* = a + b *Density Index*, where a is the density when the index is zero and b is the increase of density per unit increase of index. In wildlife counts, however, the trend of density indices on absolute density is often logarithmic (Figure 8.1), but can easily be transformed to make density linear on the index.

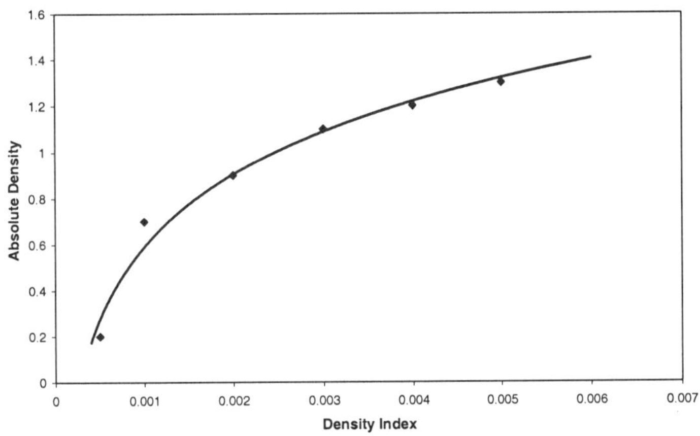

Figure 8.1: A hypothetical example of the relationship between absolute density and the density index.

As a general guideline, the linear relationship between absolute density and the transformed density index has the following form (Jachmann, 1998);

$$\log Y = a + b X \qquad or$$

$$Y = 10^{a + b X}$$

Where:

Y is density index, and
X is absolute density.

If the slope of the regression is known and the transformed regression passes through the origin (when a = 0 the index = 0), absolute density can be estimated directly from the index. However, often this relationship is not known and density indices are used in relative terms only.

Even with strictly standardised conditions, the index may not be linear on absolute density. This implies that a doubling of the index does not necessarily mean a doubling of density. Assuming a logarithmic relationship, the index will show a gradually slower increase compared with absolute density. A logarithmic transformation of the index will usually provide a near to linear relationship with absolute density.

In African wildlife management, index counts are used in a variety of ways: 1) as catch per unit effort, such as trophy animal abundance related to hunting success, 2) as direct index counts of animals, such as number of animals seen per km of road, per patrol day or per observation day at a waterhole, or 3) as index counts of animal signs, such as number of footprints or number of droppings per unit area per unit time.

With index counts, accuracy depends on strictly standardising the conditions under which the counts are done. Depending upon the type of index count, standardising all or some of the following conditions is of the utmost importance:

- **General Conditions of Counting**: The transport platform should be the same for each count. There is no point in comparing a count done with a vehicle with a similar count done on foot. Disturbance caused by a vehicle is not the same as disturbance caused by people on foot. Also, the speed at which animals are searched and counted with a vehicle is significantly different from that on foot. When using hunting success as catch per unit effort, there is no point in comparing hunts done by vehicle with hunts done on foot. Seasonal conditions and weather conditions should be the same. One cannot compare a wet-season count with a similar one in the dry season. Also, a count conducted during a cloudy day with rain should not be compared with one done during a dry spell. Water and food availability vary between seasons, and so does animal distribution. Animal behaviour will be different on a cool cloudy day with rain compared with a hot sunny day. Not only the time of year is important, but also minor considerations such as the time of day. It is not valid, for instance, to compare a count done at midday, when animals are hiding in the shade, with one done in the early morning hours, when animals are grazing or browsing. In summary, timing and transport platform should be strictly standardised.

- **Conditions of Counting Efficiency**: The number of observers and the level of proficiency of each observer should be the same for each count. For instance, assuming the same level of experience, two observers will see more animals than one observer. Also, one very experienced observer may see more animals than two less experienced observers. Using hunting success as catch per unit effort, the efficiency may vary greatly with the type of equipment used. Telescopic sights, for instance, are more efficient than open sights. In summary, numbers and skill of the observers and equipment should be standardised for catch per unit effort analysis.

- **Methodology**: The effort may be expressed in terms of unit time, such as numbers of animals seen per patrol day or hour's walk, but it may also be expressed in unit length or unit area, such as numbers of animals seen per km of road or per km^2 of study area. Here, sampling techniques follow those described in previous chapters, such as for instance the use of twice the mean perpendicular distance as the strip width.

- **Sample Conditions**: When comparing the density indices of two populations living in different areas, one should make sure that the samples are drawn from similar habitat types. An elephant population living in a rich alluvial area in one of the rift valleys will have different properties than a population living in miombo woodland. The same applies to counts of the same population, but at different times. For instance, elephants in the Luangwa Valley occupy different habitat types within the same conservation area. There would be little point in comparing an elephant count done in miombo woodland one year with a similar count done in mopane woodland the next year. Thus, when a density index is required for the entire conservation area, the sample has to be representative of the area and should be consistent between counts.

- **Disturbance**: When comparing relative densities of animals between different areas and different time periods, the disturbance caused by the survey method should be the same. The disturbance you cause depends upon the shyness of the target species in relation to the law-enforcement effort in the area. Assuming all other conditions equal, a road count of elephants in a well-protected National Park will give different results from a similar road count done in a hunting area. Thus, levels of law-enforcement effort and conditioning of the target species to the transport platform should be the same.

These conditions serve as general guidelines and may not equally apply to the index techniques listed below that are commonly used in wildlife management.

The following index techniques may provide a cost-efficient alternative to direct counts discussed in previous chapters:

1. **Hunting Success**: Relating hunting success to trophy abundance is widely used in wildlife management. With strictly standardised conditions, hunting success as a relative measure of density will provide a cost-efficient tool for an adaptive management system. There are various ways of relating hunting success to abundance; the most simple index of density is obtained by dividing the number of animals reported killed by the number of licenses issued. The resulting hunting success ratio can then be used as an index of density. At low densities, the value of the index (hunting success ratio) is low and usually linear on absolute density (Figure 8.1). Thus, at low densities the untransformed ratio provides an accurate index of abundance. At higher densities, however, the index requires transformation to make index linear on density. The major problem with this method is that there are so many factors involved. Different types of clients will vary greatly in the type of trophy they will accept, often focussing on different species. With multi-species hunts, clients may switch target animals in a single day, when accidentally running into a good trophy. A much better index of hunting success is trophy quality.

2. **Trophy Quality**: Another technique involves the time period it takes to hunt an animal with an acceptable trophy size, where the acceptable trophy size is predetermined by management according to international standards. With strictly standardised conditions, a shorter period shows that the target species is more abundant, whereas a longer period shows the opposite. Hunting quotas can then be adjusted accordingly. The same technique can be used with trophy size. When the average trophy size becomes gradually smaller, the quotas will have to be adjusted downward. This technique can be improved by combining trophy size and time required to hunt the animal. The ratio of trophy size (*m*) to time required (*days*) provides an index of trophy species abundance.

There are six widely used measuring systems for trophy quality, namely, Rowland Ward for Africa, Conseil International de Chasse (CIC) for Europe, Boone and Crocket for North America and Douglas for New Zealand and Australia. The most widely used system is that of Safari Club International (SCI), which scores all trophies with like characteristics in the same way regardless of where in the world they are found. A sixth simplified system is used by some big trophy handlers and taxidermists, in case they have a trophy mix-up. The most useful system is that of SCI. This system divides species into groups, each of which is measured in a different way, according to the nature of the trophy. There is, however, a problem with some of these measuring systems, of which SCI is aware. In the case of *bovids*, they tend to encourage hunters to shoot young males in up-coming breeding stock, since horn length reaches a maximum in young maturity. Later, horns tend to get shorter, but fatter at the base. SCI is considering changing some of the scoring systems to place more emphasis on older animals, especially in buffalo. Thus, using trophy quality as an index of animal abundance has the same problem as hunting success, because different types of clients take different types of trophy. American clients tend to go for SCI records, whereas European clients go for old trophies, while citizen hunters are less selective. It is a relatively young science that has not been analysed in any depth. Moreover, inspection of the distribution of trophy quality measurements for buffalo and kudu in Botswana (Figure 8.2), using the SCI scoring system, reveals a near to normal distribution of trophy size (Debbie Peake, pers. comm.). Thus, hunters are generally not as selective as may be expected. Also, from 1997 to 1998, there was a minor increase in the average trophy size of buffalo and kudu (Figure 8.2). This type of information is used to adjust the quotas for the next hunting season. In our example, an upward adjustment of the annual quotas for buffalo and kudu would have been justified.

The SCI system uses the following scoring system:

- *Most African antelopes with simple horns:* length and circumference of both horns are added together for the total score.
- *Spiral-horned antelopes*: length around the spiral and circumference of both horns added together as the total score.
- *Wildebeest*: tip-to-tip measurements around the outer curves and across the forehead, plus the circumference of both bosses give the total score.
- *Buffalo*: tip-to-tip measurements around the outer curves, plus the surface widths of both bosses at their widest parts.
- *Rhinos*: length and circumference of both horns give the total score.
- *Warthog, bushpig and hippo*: the length of the outer curve to the circumference at the largest place of both tusks is added.
- *Elephant*: the weight of both tusks is added to give the total score.
- *Cats*: length and width of the skull.

The system used in Zambia is not very different from the SCI system, and is provided as an example in Table 8.1.

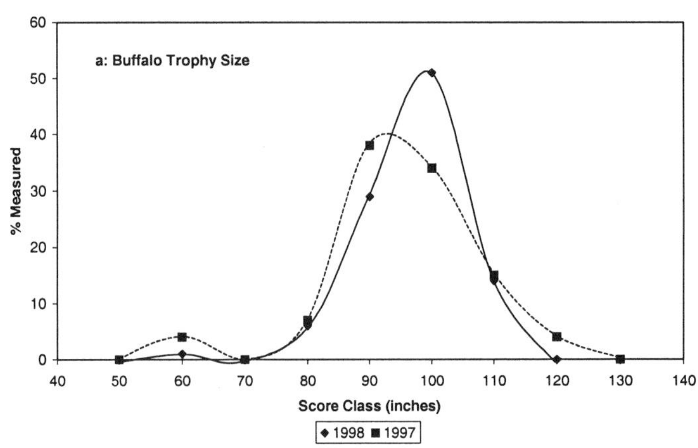

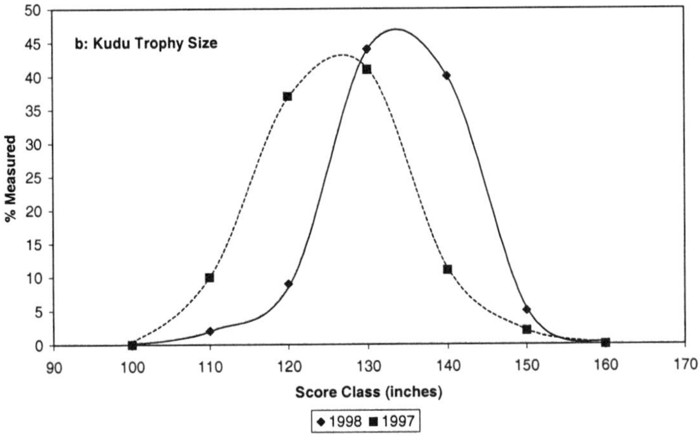

Figure 8.2: An example of the distribution of trophy quality measurements for buffalo (a) and kudu (b) in Botswana (Source: Safari Services Ltd, Maun, Botswana).

3. **Direct Index Counts**: Counts of animals per unit road, unit patrol, unit area or unit time are frequently used as indices of density. As with all index techniques, conditions will have to be strictly standardised if useable results are to be expected.

4. **Index Counts of Animal Signs**: Counts of animal signs, such as footprints and droppings, often provide a better index of density than direct index counts. Although dry-season footprints of most large African ungulates tend to be short-lived, droppings accumulate over long periods. Therefore, droppings provide an index of mean density over a period of time rather than an index of current density. This not only provides a more accurate account of relative density, but also provides useful information on the distribution of a particular animal population. A further advantage is the stationary disposition of the study objects, facilitating easy observation, while it is less critical to standardise general conditions of counting and counting efficiency. If estimates of other variables are available, such as seasonal defecation rates and decomposition rates, dropping indices can be converted to estimates of animal numbers. The density of animal signs is sampled in the same way as the density of animals, and will be discussed in detail in the next section, under indicator counts.

Table 8.1: Example of a trophy animal data sheet used in Zambia.

TROPHY ANIMAL DATA SHEET	
Dates of Hunt: from.........to.......... Area:.. Recorder:...................................... Professional Hunter:...................... Client:..	Species Hunted:................................ Firearm Used:.................................. Telescopic Sight (Yes/No):............. Transport Platform:.......................... Days Required:................................. Trophy Quality Hunt (Yes/No):....... Location of Trophy:.........................
LION	**LEOPARD**
Weight:............................. kg Body Length[1]:.................... cm Skull Length:..................... cm Skull Width:...................... cm	Weight:............................. kg Body Length[1]:...................... cm Skull Length:......................... cm Skull Width:......................... cm
BUFFALO AND WILDEBEEST	
Tip to Tip (over head):............. cm Width (buffalo):....................... cm	Tip to Tip (around boss): ....cm Circumference (wildebeest): Left[2]:..cm Right[2]:cm
ANTELOPE SPECIES[3]	
Length of Horn: Left:........ cm Right:....... cm Boss Circumference: Left[2]:........ cm Right[2]:....... cm	
1. Body length does include tail 2. Circumference is measured at the base of the boss 3. With spiral horned species (Kudu, Eland, Sitatunga, Bushbuck) horn length is measured following the spirals	

8.1 Field Example of a Patrol Index Count

The small Kasanka National Park (470 km²) is patrolled by teams of 3–4 scouts, at a rate of 5 to 8 patrols per month. During these patrols, the scouts record sightings of less common animal species. Sightings of elephants, roan antelope, sable antelope, hartebeest and waterbuck were related to the patrol effort, in terms of numbers of animals seen per patrol man-day, for 1996 to 1999. The park is surrounded by other conservation areas with small stocks of wildlife. Therefore, seasonal movements of animals across park boundaries are a common feature in Kasanka. One of the objectives of the patrol report analysis was to investigate the extent of these seasonal movements. The data for 4 years were pooled, and to minimise variation in sightings, the index was calculated for consecutive two-month periods, starting with January/February. The result is an index of animal density, in numbers seen per patrol man-day per 2 months period, showing the time of year a particular species is most abundant in the park. The index is multiplied by 1,000, to facilitate interpretation. From January to August, the index for elephants is stable between 80 and 90, corresponding to 0.08 and 0.09 elephants seen per man-day on patrol (Figure 8.3). During the late dry season and the early wet season (September – December), the index declines to approximately 60, indicating that roughly 33% of the population moves out of the park. A road dropping count in the dry season gave an estimate of 76 ± 9 elephants (Jachmann and Phiri, 1999a; *see Chapter 9*). Assuming a linear relationship between the index and actual numbers, the index of 90 (0.09) corresponds to approximately 76 ± 9 elephants. Similar patrol report indices showed that sable and hartebeest use the park on a seasonal basis, with highest densities during the late dry season (Figure 8.4). Both roan and waterbuck occur in low densities year-round, with minor peaks in the early and late dry season (Figs. 8.4 and 8.5).

If we assume the same relationship between the index and actual numbers as for elephants, we can now convert the index for the other 4 species into approximate numbers. The results can then be compared with estimates obtained from an aerial sample survey conducted on 13 and 14 October 1999 (Table 8.2).

The aerial sample count gave imprecise results due to low densities and uneven distribution (Table 8.2). Sable antelope start arriving in the park in September, and reach their highest densities in December (Figure 8.4). During the aerial survey in October, no sable were seen. Roan, hartebeest and waterbuck, however, live year-round in the park, and estimates from the aerial survey can be compared with numbers converted from the index at the time of the survey (Table 8.2). In spite of low densities, and therefore the imprecise results from the aerial sample count, the converted numbers for roan, hartebeest and waterbuck compare well with those from the aerial count, and are not significantly different. This confirms linearity at low values of the index (Table 8.2).

In conclusion, we may state that patrol reports provide an excellent and cheap option in situations where alternative methods give imprecise results, and above all, are very expensive.

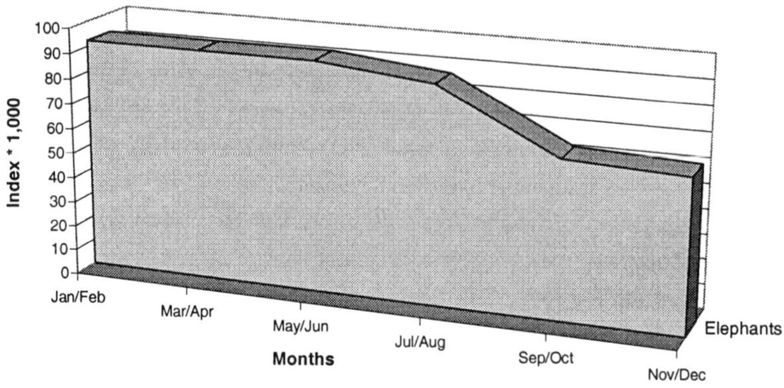

Figure 8.3: Index for elephant sightings on patrol, in numbers seen per patrol man-day per 2 months (x 1,000), from 1996 to 1999.

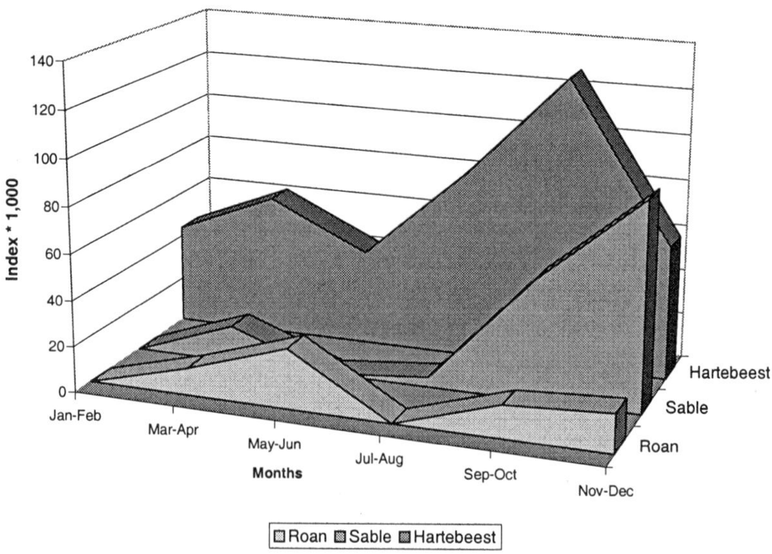

Figure 8.4: Index for roan, sable and hartebeest sightings on patrol, in numbers seen per patrol man-day per 2 months (x 1,000), from 1996 to 1999.

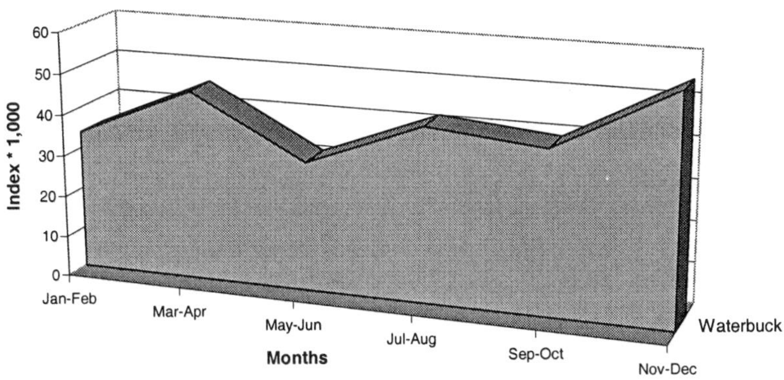

Figure 8.5: Index for waterbuck sightings on patrol, in numbers seen per patrol man-day per 2 months (x 1,000), from 1996 to 1999.

Table 8.2: Patrol index converted to numbers, using the result of an elephant dropping count, compared with the results of an aerial sample survey conducted in October 1999.

	INDEX COUNT			AERIAL SAMPLE COUNT	
Species	**Index (x 1,000)**	**Numbers (Converted)**	**SE**	**Estimate**	**SE**
Elephants	90	76*	9	27	19
Roan	26	22	3	54	56
Sable	80	68	8	-	-
Hartebeest	122	103	12	104	49
Waterbuck	39	33	4	25	18

* *Road Dropping Count.*

8.2 Monitoring of Sex and Age Ratios, and Juvenile Mortality

Trends in wildlife populations are determined by the relative rates of recruitment and immigration, versus mortality and emigration. When animal movement is restricted, the sex and age ratios of large mammal populations, in combination with juvenile survival rates, may serve as useful indices of recruitment and of changes in population growth. This technique will provide valuable information for managers in conservation areas that are fenced, or where animal movement is restricted by landscape features or cultivation, and where mortality through illegal hunting is low. Detailed information on population dynamics will provide a better insight into population fluctuations than the mere assessment of abundance. The main disadvantage is that sampling of sex and age ratios and juvenile mortality is time consuming and therefore expensive. These data must be collected over several years at the same time as other methods are used to estimate abundance. The Kruger National Park is the only place where the area is sampled for sex and age ratios and juvenile mortality, corrected for rainfall, relating population dynamics parameters to population trends obtained through line-transect methodology (Mason, 1990). Because this technique requires a large and well-established research team, supported by ample funds, we will not go into further detail.

PART IIIb

INDICATOR TECHNIQUES

With large African mammals, only two indicator techniques are commonly used; e.g. dropping counts and track counts. Dropping counts have been used for a variety of species, whereas track counts, mainly footprint measurements, have been reported for elephants, rhino and duiker. The technique of using faecal deposits for estimating density, distribution and trend in large game populations originated in the United States in the 1930s (Neff, 1968). Pellet-group counts were used to estimate abundance of several species of deer, mountain elk, moose and Barbary sheep (Neff, 1968). It was not until the mid 1960s that the technique was introduced in Africa, to estimate changes in the abundance of elephants in the Budongo Forest in Uganda (Laws *et al.*, 1975). Since the 1960s, the technique has been widely applied to estimate elephant abundance, but it was not until the mid-1980s that the use of dung began to gain popularity for studying other species. In the 1980s, dung was used to estimate numbers and distribution of a small rhino population in the Lower Shire Valley in Malawi (Jachmann, 1984a) and the abundance of duiker in the Ituri Forest in the Zaïre (Koster and Hart, 1988). During the 1990s, its application was expanded to bushbuck, buffalo and mountain gorilla (in addition to duiker and elephant) in Rwanda (Plumptre and Harris, 1995). Over the past 30 years, however, most applications concern elephant populations. In Africa, the technique was used to estimate numbers and densities of elephant populations in Malawi (Jachmann and Bell, 1979,1984; Jachmann, 1984b), Ivory Coast (Merz, 1986), Ghana (Short, 1983), Burkina Faso (Jachmann, 1988b, 1989 & 1991a), Central African Republic (Fay, 1991), Congo (Fay and Agnagna, 1991), Zaïre (Alers *et al.*, 1992) and Gabon (Barnes *et al.*, 1991, 1995 & 1997a). The technique is suitable for elephants, first, because elephant droppings remain readily visible for extended periods, and second, because of the availability of information on variables that are difficult to measure, such as defecation rates and faecal decomposition. There exists ample literature on the use of faecal material to estimate animal abundance, including a review (Neff, 1968) and several field guides (Barnes and Jensen, 1987; Dawson and Dekker, 1992). Also, because dung is used more and more frequently to count wildlife, particularly in forests, the technique has rapidly evolved and has changed beyond recognition. The first chapter in this section (Chapter 9) briefly summarises the various sampling techniques for estimating elephant abundance under different field conditions, while the second chapter (Chapter 10) discusses the use of footprint and dropping measurements to estimate numbers and distribution of small isolated rhino and elephant populations.

Chapter 9

DROPPING COUNTS

The use of dropping counts in ecological studies is not always accepted by laymen as a valid method. The lack of credence for the method stems from the fact that laymen and many professionals alike feel that counting excrement is a trivial and banal activity, in addition to the fact that no live animals need to be observed, which makes it even more suspicious. However, droppings are available in large numbers and often accumulate over several months, giving the advantage that any information extracted from dung shows less variation than direct counts, and relates to animal distribution over the preceding months. A direct count, however, records the instantaneous distribution, usually with small sample sizes, and is therefore more prone to sampling error (Jachmann, 1991a). In the past, dropping counts have been shown to be superior to other techniques, providing an accurate means of, for instance, assessing elephant densities (Jachmann, 1991a; Dawson, 1990; Sukumar *et al.*, 1991; Barnes, in press). On the Nazinga Game Ranch in Burkina Faso, different methods of counting elephants were tested (Jachmann, 1991a). We assumed that the aerial total count was close to the true number of elephants on the ranch. Dropping counts gave both the most accurate and most precise estimates (Table 9.1).

Table 9.1: Elephant population estimates (dry season) in the Nazinga Game Ranch, Burkina Faso (Jachmann, 1991a).

Method	Year	Estimate	95% CL	Estimator
Aerial Total Count	1989	366		Complete Count
Aerial Sample Count	1988	610	0 – 1,270	Fixed-width Strips
Line-transect Count	1987	487	210 – 774	Fourier Series
Line-transect Count	1988	306	0 – 952	Modified Hayne
Vehicle Count	1988	293	0 – 728	Mean Perpendicular Distance
Droppings	1987	396	323 – 469	Droppings Accumulation
Droppings	1987	353	276 – 430	Steady State Assumption
Droppings*	1987	420	0 – 910	Droppings Accumulation

*Wet-season road count

Because most of the dropping counts described in the literature deal with elephants, this chapter will focus on their use for estimating elephant abundance. However, the same technique, with minor modifications, can be applied to most wild animals.

Dropping densities can be used as an index of abundance which is adequate for most management purposes. However, as pointed out by Barnes (1993, 1995), much of wildlife management is politics, and dung carries little weight with politicians. Also, the strong effect of rainfall on dung decay, combined with short-term and long-term changes in rainfall, necessitate converting dung to numbers (Barnes *et al.*, 1997b).

The basic concept of dropping counts is straightforward. The number of dung-piles counted in the sample area has to be converted to numbers of elephants. This requires two variables; the defecation rate, or the average number of dung-piles produced per elephant per time unit, and the decay rate, or the average time period it takes for dung-piles to decompose and become unrecognisable to observers. Both variables have a strong seasonal dependency. With dropping counts, the sample design primarily depends on the type of habitat, and the accuracy and precision required. The first differentiation is between woodland savannas and moist evergreen forest. In woodland savannas, fixed-width transects, plots and line transects can be used as the sample units, whereas in forest the line-transect technique is the only practical option (Barnes and Jensen, 1987). In forest it is extremely labour intensive to clear the dense undergrowth for fixed-width transects or plots, while revisits of permanent sample units are usually impractical (Barnes, 1993). In contrast, the variable visibility profile inherent to the line-transect technique allows application under most conditions. Transects may be visited once if one of these assumptions holds: (i) the number of droppings on the ground is in a steady state, with deposition equalling decomposition (McClanahan, 1986), or (ii) rainfall during the preceding months is used as a predictor variable to determine dropping densities (Barnes *et al.*, 1994; Barnes *et al.*, 1997b). In dense forest, the one-off dropping count method is the only practical technique available to estimate numbers of elephants or other animals, but in woodland savannas this method is just one of several alternatives. Although the dropping method has been shown to produce accurate results under different field conditions, with transects and plots the method is labour intensive, and when applied to large savanna areas may be more costly than alternative techniques. A cheap and valid alternative is to count droppings on roads. This technique can be used in areas with an extensive road system that constitutes a representative and therefore unbiased sample of the study area.

After a brief explanation of the theoretical concept, sample design and analysis of dropping counts in woodland savanna will be discussed. This will be followed by dropping counts in forest. The literature documents many examples of dropping counts conducted in woodland savanna and forest, but very few road counts. Therefore, the application of road counts will be illustrated with several examples from the field.

9.1 Theoretical Concept

Let us assume that the rate of decay is negligible, so that droppings accumulate during the study period. *E* elephants deposit dung-piles at a rate of *D* droppings per elephant per day, which accumulate over a period of *t* days, resulting in a total of *Y* droppings in the study area at time *t*,

where ($Y_t = E \times t \times D$),

$$E = \frac{Y_t}{t \times D} \quad (1)$$

This simplified concept is only valid under conditions where dropping density is known at the beginning of the accumulation period *t*, and when no droppings are lost through decomposition. Although Jachmann and Bell (1979, 1984) reported that in Kasungu National Park in Malawi, there appeared to be no effective decomposition of droppings during a period of 2 – 3 months in the late dry season, most other researchers working in woodland savanna reported dry-season decay rates of between 10% and 35% per month (Wing and Buss, 1970; Weir, 1971; Coe, 1977; Jachmann, 1988a, 1989). Because rainfall affects dung decay (White, 1995; Barnes *et al.*, 1994; Barnes *et al.*, 1997b), wet-season decay rates are substantially higher (Jachmann and Bell, 1984; Jachmann, 1988a, 1989). Due to high rainfall and short dry periods, rates of decay are higher in forest than in woodland savanna, and also show more variation in time and between areas. In forest, the concept of decay has received much attention, and recently new methods have been proposed which do not require the steady-state assumption (Plumptre and Harris, 1995), nor studies of dung decay (Barnes *et al.*, 1997b).

Assuming a constant mean daily rate of decay (*r*) that depends on rainfall, temperature and humidity, and will therefore vary between seasons and locations, the concept changes. Imagine *E* elephants deposit dung-piles at a rate of *D* droppings per elephant per day for *t* days. A proportion of the droppings (*r*) decomposes per day for *t* days, resulting in *Y* droppings at time *t*.

Therefore: $Y_t = E \times D \times t - E \times D \times t \times r$,

$$E = \frac{Y_t}{D \times t(1 - r)} \quad (D>r) \quad (2)$$

The main problem with this approach is the use of a constant rate of decay, while decay of elephant dung may follow a reverse sigmoid curve, where a period of slow decay precedes the exponential phase (Barnes, 1993; White, 1995; Plumptre and Harris, 1995). The use of a mean constant daily rate will result in underestimates, due to a moderately negative bias.

Often the use of permanent strips or plots is impractical because of the remoteness of the area or the amount of labour involved. In such cases, the steady-state assumption can be applied halfway through the dry season in woodland savanna (McClanahan, 1986), and in forest 2 months into the dry season (Barnes *et al.*, 1997b). During these periods, the deposition of droppings is assumed to be equal to the decomposition of droppings. Although the steady state assumption does not always hold (Barnes, 1993; Barnes *et al.*, 1997b), the method has shown to produce accurate results in savanna woodland (Jachmann, 1991a). The formulae for estimating elephant abundance with the steady-state assumption are shown in Box 9.1.

Box 9.1: Formulae for estimating elephant numbers, using the steady-state assumption.

The change in the quantity of droppings with time is equated as follows (McClanahan, 1986):

$$\frac{dY}{dt} = E \times D - Y \times r \quad (3)$$

Assuming a steady state, $dY/dt = 0$ and $E \times D = Y \times r$ or,

$$E = \frac{Y \times r}{D} \quad (4)$$

Where the rate of decay (r) is expressed as $ln2/t_{½}$ or $0.693/t_{½}$, where $t_{½}$ is the time (in days) at which half of the original droppings are unrecognisable, using a year-round decay rate. Equation (5) can now be completed as follows:

$$E = \frac{Y \times (0.693 / t_{½})}{D} \quad (5)$$

Calculation of variance is as follows (Goodman, 1984, quoted in Barnes, 1993):

$$Var(Y \times t_{½}) = Var(Y) \times Var(t_{½}) + Y^2 \times Var(t_{½}) + t_{½}^2 \times Var(Y)$$

$$Var(E) = Var(D) \frac{(Y \times t_{½})^2}{D^4} + \frac{Var(Y \times t_{½})}{D^2}$$

The standard error is given by: $SE(E) = \sqrt{Var(E)}$

The 95% CL are given by: $t \times SE(E)$

9.2 Dropping Counts in Woodland Savanna

Sample design in savanna woodland may follow any of the sampling procedures described in previous chapters, such as temporary or permanent fixed-width strips or plots, or line-transect methodology. Permanent sample units may be used for long-term studies in areas that are easily accessible. Transects, strips or plots may follow a stratified random design, when sufficient information is available on the population under study, or a systematic design, facilitating analysis of spatial and temporal distribution. Before we continue with analysis of dropping counts, two important variables will have to be discussed: the defecation rate and the decay rate.

9.2.1 Defecation Rate

The defecation rate, or the number of dung-piles produced per elephant per day, varies by habitat and by season. Therefore, estimates from woodland savannas cannot be used for forest elephants, and dry-season estimates cannot be used for wet-season studies. Obtaining seasonal defecation rates for a particular population is a tedious task and requires following elephants on foot for many hours. The exercise is not only labour intensive, but also hazardous. If there is not enough time and/or money to obtain this information, you will have to use data from the literature.

As pointed out by Barnes (1993), with defecation rates as well as decay rates, it is useful to distinguish between the proximate factors and the ultimate factors that determine observed differences between locations and seasons. The proximate factors governing defecation rate are soil type, habitat and availability of grass in relation to browse, since these determine food quality and availability. An increased intake of green grass will give an increased throughput and thus a higher defecation rate. The ultimate factor is rainfall, determining the quantity of grass. Based on this, a universal theory of defecation was proposed (Barnes, 1993), with rainfall as the predictor variable. However, before we can construct a model predicting defecation rate, a large body of information will have to be collected at many sites throughout the continent. Also, it is probably safe to say that independent of the location, dry-season defecation rates for elephants in woodland savannas only show minor differences from one area to the next (Table 9.2). Thus, an average rate of 15.0 ± 1.0 dung-piles per elephant per day seems a realistic average rate for studies conducted during the dry season. There are a few more defecation rates available in the literature. However, some are measurements for either males or females, but not combined (Barnes, 1982; Guy, 1975), or they are based on few hours of observation (Wyatt and Eltringham, 1974).

Because rainfall varies from one area to the next, wet-season defecation rates also show more variation than dry-season ones. The scanty information provided in the literature suggests an average rate of approximately 25 ± 2 droppings per elephant per day for the wet season (Table 9.2).

Table 9.2: Seasonal defecation rates for elephants in woodland savannas (droppings/24 hours).

Defecation Rate	Habitat	Source
Dry Season		
15.5	Captive elephants	Coe (1972)
15.7	Miombo woodland	Jachmann & Bell (1984)
14.1	Mixed woodland	Jachmann (1988a, 1989)
15.1 ± 0.7		

Wet Season		
23.6 27.2	Miombo woodland Mixed Woodland	Jachmann & Bell (1984) Jachmann (1988a, 1989)
25.4 ± 1.8		

9.2.2 Decay Rate

In the literature, very little information exists on decay rates of elephant droppings in woodland savanna. The only recent data available came from a study conducted on the Nazinga Game Ranch, in Burkina Faso (Jachmann, 1988a, 1989 & 1991a). Here the decay rate was estimated as 0.59 % per day, or *r* is 0.006 per day, with a half-life value ($t\frac{1}{2}$) of 82.7 days. This decay rate was similar to a $t\frac{1}{2}$ of 80 days, observed in Queen Elizabeth National Park in Uganda (Wing and Buss, 1970).

Decay rates for a specific location can be measured by following a large cohort of fresh droppings (>50), deposited in various locations in proportion to the size of the various habitat types in the study area. The state of the droppings will have to be observed until all of the sample has become unrecognisable. The definition of the term "unrecognisable" is different for the forest than for woodland savanna. In the forest, a disintegrated dung-pile, forming an amorphous flat mass, may not be detected at distances of more than 2 metres, whereas a dung-pile in the same state, deposited in woodland savanna, may be detected at much larger distances. The methods used to calculate decay rates are described by Barnes and Barnes (1992). As mentioned before, decay rate most likely follows a reverse sigmoid curve, which means that a period of slow decay precedes the exponential phase when decay is constant (Dawson, 1990; Barnes, 1993; White, 1995, Barnes *et al.*, 1997b). Here, the problem is to replace the constant rate with a model for a reverse sigmoid curve, without sufficient information on the exact shape of the curve under different conditions. Firstly, this would be premature and secondly, this would unnecessarily complicate analysis. A simpler method is to approximate decay in woodland savanna by a negative exponential model, assuming a mean constant daily rate (Barnes and Jensen, 1987; Barnes, 1993). If decay really follows a sigmoid curve, this simplified approach will merely result in a small negative bias in the final estimate.

The negative exponential model has the following form:

$$N_t = N_0 e^{-rt}$$

Where:

N_0 is the initial number of droppings deposited at various locations,
N_t is the number of droppings recognisable after *t* days, and
r is the daily rate of decay.

Proximate factors governing decay rate are trampling by large animals, rummaging for seeds, termites, dung beetles, fungi and bacteria (Barnes, 1993). Ultimate factors are rainfall, humidity and temperature. As with defecation rates, a universal theory of decay, with the ultimate factors as predictor variables would be useful (Barnes, 1993). However, the very fact that in the past the majority of dung counters have used the limited information available in the literature, instead of conducting their own experiments, does not provide much hope for the massive amount of data required to construct such a universal theory for woodland savanna. For forest, however, considerable progress has been made (*see section 9.3*).

Dry-season decay rates in woodland savannas may not greatly differ from one location to the next. For the dry season, an average t ½ of 80 ± 5 days can be used, and for the wet season an average t ½ of 49.6 ± 3 days (Jachmann, 1991a).

9.2.3 Analysis

With temporary fixed-width strips and plots, or with line-transects, dry-season dropping counts can be analysed assuming a steady state (see Box 9.1). The procedure of estimating the variance requires calculating the variance of a product and a ratio (Barnes, 1993). Other methods for estimating variance exist, but they are more complicated and require additional work.

Analysis of dropping counts conducted in woodland savanna using permanent transects or plots is shown in Box 9.2.

Box 9.2: Formulae for estimate and precision of dropping counts, using permanent sample units.

The first step is to estimate the number of base-line droppings for *n* sample units (ΣB_0). The count has to be repeated at time *t* (1 to 3 months later), giving the total of the base-line droppings and the accumulated droppings at time *t* (ΣB_t).

The droppings that accumulated over the study period can be calculated by deducting the base-line from the numbers of droppings at time *t* for each of *n* sample units:

$$\sum_{i=1}^{n} Y_i = \sum_{i=1}^{n} (B_{i0} - B_{it})$$

The next step is to calculate the size of the sample area (A in km²). Equation (2) can now be changed as follows:

$$E = \frac{\sum_{i=1}^{n} Y_i}{(1 - t \times r)t \times D \times A} \quad (D>r) \quad (6)$$

Where:

E is elephant density in the sample area (elephants/km²),
n is the number of sample units,
$\sum Y_i$ is the number of droppings accumulated in the sample area during the study period (t),
r is the mean daily rate of decay,
t is the accumulation period (days),
D is defecation rate (dung-piles produced per elephant per day), and
A is size of sample area (km²).

The precision of the estimate can be approximated by the sum of the coefficients of variance (CV) of each of the variables (conventional error analysis). The coefficient of variance is calculated by dividing the Standard Error by the Mean ($CV = SE/Mean$). The expression for equation (6) is:

$$CV^2(E) = CV^2(Y) + CV^2(r) + CV^2(D) + CV^2(A) \quad (7)$$

The mean and standard error of the number of droppings in the sample area ($\sum Y$) can be calculated by taking the mean of all sample units and the variance between sample units.

The standard error of the decay rate (r) can be approximated at 6% (0.006 ± 0.0004, *see section 9.2.2*), and the standard error of the defecation rate (D) at 6.7% (15.0 ± 1.0, *see Table 9.2*). The error in measuring the size of the sample area (A) depends on the method. For instance, with a ruler and a map, the error is determined by the smallest unit on the ruler and the error in the map, and may be as much as 10%. When the exact size of the sample area is known, A is a constant, otherwise, measurement errors are treated in the same way as sampling errors.

9.3 Dropping Counts in Forests

Sample design for dropping counts in forests follows line-transect methodology. Procedures are exactly the same as those described for direct counts (*see Chapter 4*). The notable difference with direct counts is the necessity to stratify the study area for accurate results, accounting for human population density (Barnes, 1993; Barnes *et al.*, 1991). Human population density is a measure of general human disturbance, determining elephant numbers and distribution in the forest (Barnes, 1993). The human factor results in a gradually increasing dropping density with an increasing distance from roads or human settlements (Barnes, 1993), and is probably a major source of bias leading to underestimates of forest elephant populations. This is especially the case when transects do not penetrate deeply enough into large and remote areas of forest. Barnes (1993) suggested placement of a base-line perpendicular to roads or rivers, with random transects along the base-line. The data would then be used to construct a curve relating dropping density to distance away from the road or river, which in turn can be used to stratify the study area.

In the forests of Gabon, dung-pile density could be explained by distance from the nearest road (X_{rd}) and the human density (*STRATUM* is high or low human density) by the following equation (Barnes *et al.*, 1991; Barnes *et al.*, 1997b):

$$\sqrt{Yt} = a + b\log(X_{rd}) + d\text{STRATUM}$$

The population estimate can then be obtained by combining this figure with a GIS (Barnes *et al.*, 1997a).

9.3.1 Defecation Rate

In the forest, the main problem is following elephants on foot for long hours in dense undergrowth. This is labour intensive and above all, more dangerous than in savanna woodland. For many years, the estimate of 17.0 droppings per elephant per day (Wing and Buss, 1970) was used by dung counters. However, as pointed out by Barnes (in press), their defecation rate was measured in the woodland savanna of Queen Elizabeth National Park, outside the Budongo Forest, where most of their research was concentrated. Compared to more recent measurements in the Ivory Coast of 18.0 (Merz, 1986) and 19.8 droppings per day in Cameroon (Tchamba, 1992), the estimate of 17.0 is probably on the low side. Because estimating defecation rates in the forest is potentially dangerous, dung counters can use an average rate of 19.0 ± 1.0 droppings per elephant per day. Although the use of defecation rates from other sites may introduce a potential source of error, this can never be too serious. The most accurate result obtained from a dropping count is 7.04 elephants in a small study area in Parc National des Vulcans in Rwanda, while the actual number was 7 (Plumptre and Harris, 1995). The method avoids the steady-state assumption, but uses Wing and Buss's (1970) estimate for the defecation rate.

9.3.2 Decay Rate and Analysis

An important source of error in dung counts conducted in the forest is the assumption that the system is in a steady state (McClanahan, 1986; Barnes, in press). In woodland savanna the system may be in equilibrium during part of the dry season. In forest, however, the system is rarely stable (Barnes *et al.*, 1995, in press; Plumptre and Harris, 1995). Alternative models have been developed, avoiding the steady-state assumption. Plumptre and Harris (1995) developed a matrix model, in which the actual dung counted in the transects was corrected for monthly decay. The major disadvantage of this model is that monthly decay rates have to be measured. A more sophisticated model relates rainfall in the two months preceding the count to dung density, so avoiding the steady-state assumption and the tedious task of monitoring dung decay (Barnes *et al.*, 1997b). Their research resulted in the following equation:

$$Y_t = 1{,}071.09 - 0.92RAIN_{t-1} - 0.59RAIN_{t-2}$$

where:

Y_t is dung density,
$RAIN_{t-1}$ is rainfall in the month preceding the count,
$RAIN_{t-2}$ is rainfall in the month before that.

Adding the dropping density in the preceding month gave a small, but significant improvement (Barnes *et al.*, 1997b). However, including this extra term to the equation requires two counts, something we were trying to avoid when counting dung in inaccessible forest areas. The example provided by Barnes *et al.* (1997b) is shown in Box 9.3.

Box 9.3: Example of the use of the rainfall model, developed by Barnes *et al.* (1997b).

> In September, the mean dropping density in a forest is estimated at 269 per km². The rainfall during August was 62 mm and during July 161 mm.
>
> If the elephant density were one per km², dropping density (Y_t) would be:
>
> 1,071.09 – 0.92 x 62 – 0.59 x 161 = 919.05 per km² (*see equation above*),
>
> However, dropping density was 269 per km², and therefore:
>
> The density of elephants in the study area must be 269/919.05 = 0.293 per km².

If information on rainfall is not available, and when dung decay has not been measured, the steady-state assumption may be used after the rains have stopped for 2 months, using the rate of 0.024 per day (Barnes and Jensen, 1987).

9.4 Dropping Counts on Roads

In woodland savanna, dropping counts on roads provide an attractive, quick and cheap alternative to conventional dropping counts. An extensive road system must exist that is more or less representative of the different habitat types and preferably cuts across the main drainage system(s). Elephant density was estimated by road counts in the Nkhotakota Game Reserve in Malawi (Jachmann, 1984b), in the Gourma region of Mali

(Jachmann, 1991b), and in the Nazinga Game Ranch, Burkina Faso (Jachmann, 1991a). The wet-season road count of droppings in the Nazinga Game Ranch gave an estimate of 420 elephants, which was only 6% higher than the aerial total count (Table 9.1). Although the estimate was accurate, it was imprecise (SE is 59.52%).

After a brief discussion on the validity of the method, sample design and analysis will be discussed, illustrated with two examples from the field.

9.4.1 Validity of the Method

Prior to the dropping count in the Nkhotakota Game Reserve in Malawi (Jachmann, 1984b), the validity of the method was assessed by applying it to an area of known elephant density in the Kasungu National Park. The elephant density in the high-density stratum of Kasungu National Park was estimated by aerial survey at approximately 3.1 ± 0.5 elephants per km² (Bell and Jachmann, unpublished aerial survey data). In this stratum, a 19.4 km long track, with a mean width of 3.2 m, was cleared of droppings. Then it was sampled 7 times over a 40-day period. The elephant density, estimated from each of the 7 dropping counts, follows a negative logarithmic curve, approaching a stable value after approximately 30 days (Figure 9.1). This period will be referred to as the minimum accumulation period, after which sampling can be carried out with a minimum of sampling error.

After 40 days, elephant density from the track count was estimated as 3.6 ± 0.8 elephants/km² (Jachmann, 1984b), which is not significantly different from the density obtained from the aerial sample count (3.1 ± 0.5). The fact that elephant density, estimated from 7 consecutive dropping counts, declines towards a stable value within a period of about one month can be explained as follows: with the exception of adult bulls, elephants travel in family units. Usually several closely related family units, called a kin-group, travel together in a broad unit line, separated by only a few hundred meters of woodland (Jachmann, 1984b & c). It is obvious that the first dropping count, after the road had been cleared of droppings, is either biased towards too many or too few droppings, due to the fact that a group of elephants either misses or hits the track. In the course of time, droppings accumulate and a sudden increase in numbers will be a decreasing relative proportion, eventually resulting in a more or less stable number of droppings. Thus, the longer the accumulation period, the lower the sampling error.

First, the experiment shows that dropping counts on roads may be a valid method to estimate elephant density, and second, when using the technique of droppings accumulation after clearing the road, a minimum accumulation period of 30 days is required to minimise sampling error. However, the dropping road count assumes that elephants are distributed independently of roads, i.e. the roads are a representative random sample of the study area.

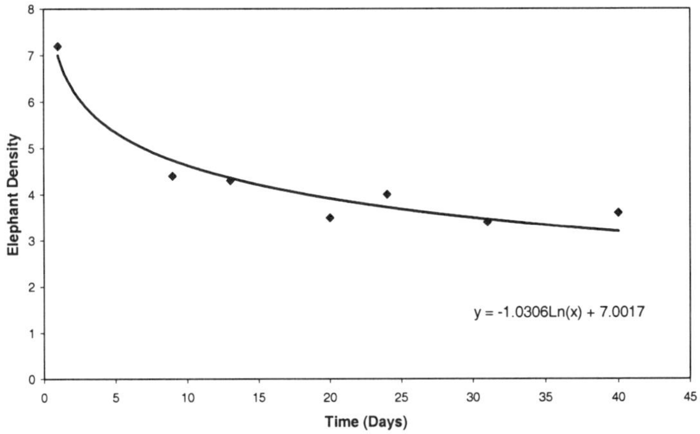

Figure 9.1: Elephant density, estimated from 7 dropping counts, carried out over a 40-day period on a 19.4 km long track in Kasungu National Park, Malawi (Jachmann, 1984b).

During the experiment in Kasungu National Park, all of the droppings deposited during the accumulation period of 40 days were flattened, but none showed any signs of decay. During the dry season, when roads are frequently used, droppings are flattened by passing vehicles, resulting in negligible termite activity and long-term survival of droppings. Also, flattened droppings remain visible on roads for a couple of months. During the wet season, however, droppings are rapidly washed away. Thus, dropping decay on roads appears to be a special case: decay is low during most of the dry season, but rapid in the wet season. Therefore, a road dropping count conducted during the dry season, after a short period of accumulation (approximately one month) after the roads have been cleared, can be analysed assuming zero decay. A count conducted after a long period of accumulation, without first clearing the roads, may be analysed with the steady-state assumption and a year-round decay rate.

9.4.2 Sample Design and Analysis

With dropping counts on roads or tracks, the sample design depends on the road-system. In areas with an extensive road system, such as on the Nazinga Game Ranch in Burkina Faso, roads can be selected to include each habitat type in proportion to its surface area. If for instance 50% of the study area consists of miombo woodland, then 50% of the roads or tracks sampled should pass through miombo woodland. Also, roads selected for counting should preferably run along density gradients of the target species, which means at 90° angles to the drainage system. Frequently, the road system is constructed along areas with high animal densities, such as along rivers or in valleys, which leads to overestimates of abundance.

Roads or tracks are artificial, void of vegetation, on the one hand a potential source of disturbance, but on the other hand sometimes providing safety. They can be used by elephants to move from one place to another. Because roads lack vegetation, they may be under-utilised, but they may also be over-utilised when vegetation along the road is preferred. Roads may be avoided by breeding herds, but used by adult males. Some of these biases are positive, leading to overestimates, while others are negative, leading to underestimates. However, since dropping counts on roads can provide accurate results (Jachmann, 1991a), most of these biases are small and usually balance out.

The length of roads counted is shown by the odometer of the vehicle, while the average width requires a number of measurements at fixed intervals. Sometimes elephants defecate while walking across the road, which makes it difficult to determine whether the centre of the dropping is on the road and the dropping should therefore be counted or falls outside and should be ignored. Fortunately, this source of bias is small (< 2%).

Analysis follows the same procedures as other dropping count techniques. To illustrate this, the following field examples will be analysed in different ways.

9.4.3 Field Examples

The Central Luangwa Valley: From June to August 1994, a road count of elephant droppings was done in 5 conservation areas in the central Luangwa Valley; e.g. South Luangwa National Park, Luambe National Park, and Lupande, Lumimba and Munyamadzi Game Management Areas. There are very few roads in these conservation areas, and several run alongside major rivers. Thus, the results of the dropping counts were expected to be biased towards too many elephants. Nevertheless, the count was considered an experiment, and results were compared with estimates from an aerial survey carried out in October 1994. Prior to the count, all roads were cleared of droppings. The accumulation period, or the period between clearing the roads and the count, varied from 32 to 34 days. It was assumed that dropping decay during this period was negligible. The average width of the roads was 3.1 ± 0.2 m. The error in measuring the length of road was arbitrarily taken as approximately 2%, the error margin in the size of the survey block as 15%, and the error in the number of droppings counted at 2%. The few roads in the valley resulted in a low sampling intensity or area coverage, ranging from about 0.008% to 0.02%. Length of road sampled, size of sample area, size of survey block and numbers of droppings counted in each area, together with the measurement errors, are shown in Table 9.3.

Table 9.3: Length of road sampled, size of sample area, size of survey block, and number of droppings counted in each conservation area, including measurement errors.

Area	Length of Road (km)	Sample Area (km^2)	Survey Block (km^2)	# Droppings
South Luangwa NP	202 ± 4.0	0.6262 ± 0.0423	8,000 ± 1,200	277 ± 5
Luambe NP	22 ± 0.4	0.0682 ± 0.0019	110 ± 16	28 ± 1
Lupande GMA	74 ± 1.5	0.2294 ± 0.0066	800 ± 120	139 ± 3
Lumimba GMA	182 ± 3.6	0.5642 ± 0.0160	2,500 ± 375	42 ± 1
Munyamadzi GMA	98 ± 2.0	0.3038 ± 0.0088	1,000 ± 150	103 ± 2

Analysis of population estimates uses equation 6 (*see section 9.2.3*), omitting decay rate, and estimation of variance uses equation 7, including the measurement error for the size of the survey block. Estimation of variance uses measurement errors in exactly the same way as standard errors.

Estimation of population numbers for South Luangwa National Park is as follows:

1. The area sampled (A) is 202 km (length of road) x 0.0031 km (width of road) is 0.6262 km^2 (Table 9.3),

2. The number of droppings in the whole park; Y = (8,000 km^2(size of survey block)/0.6262 km^2(size of sample area)) x 277 droppings (number of droppings counted) = 3,538,806

3. The number of elephants is 3,538,806/32 (accumulation period in days) x 15.0 (defecation rate in dung-piles per elephant per day) = 7,372 (*E*).

Approximation of the population variance is as follows:

1. $CV^2(E) = (5/277)^2 + (0.0423/0.6262)^2 + (1200/8000)^2 + (1/15)^2 = 0.027726$,

2. $CV(E) = \sqrt{0.027726} = 0.1665$,

3. $SE(E) = 0.1665 \times 7{,}372 = 1{,}228$,

4. 95% CL = 1.96 x 1,228 = 2,407

Result: 7,372 ± 1,228 elephants (95% CL 4,965 – 9,779). There are definitely at least 4,965 elephants in South Luangwa National Park, probably as many as 9,779, while the true population size is possibly around 7,372.

Table 9.4: Comparison of elephant population estimates from road dropping counts with those from an aerial sample count, both carried out during the dry season of 1994.

	Dropping Count	**Aerial Sample Count**
Area	**Population Estimate (SE)**	**Population Estimate (SE)**
South Luangwa NP	7,372 ± 1,228	7,197 ± 1,201
Luambe NP	94 ± 16	54 ± 45
Lupande GMA	950 ± 158	829 ± 309
Lumimba GMA	388 ± 65	299 ± 164
Munyamadzi GMA	685 ± 114	459 ± 170

Because part of the roads sampled for droppings ran along the main rivers, elephant population estimates from the dropping counts were consistently higher than those from the aerial sample survey (Table 9.4). However, none of the estimates from the dropping counts are significantly different from those obtained by aerial survey, and we may conclude that even under conditions of low sampling intensity, with roads that run along high density areas, a road dropping count is a rapid and sound alternative to other survey methods for estimating numbers of elephants.

The Kasanka National Park: A small population of elephants of unknown size is found in the Kasanka National Park, which is briefly described in section 8.1. The park has an excellent road-system, and it was decided to carry out two road dropping counts. To obtain an estimate of the number of elephants using the park year-round, the roads were first sampled in September 1999. Droppings were assumed to have accumulated since the last rains in early May. A total of 180.1 ± 0.2 km of road was counted for droppings. The average width of the road was 2.2 ± 0.2 m. The number of droppings counted was 54 ± 2. The dry-season defecation rate used is 15.0 ± 1.0 dung-piles per elephant per day, with a year-round decay rate (t ½) of 49.6 ± 3 days and a dry-season decay rate of 0.006 ± 0.0004 per day (Jachmann, 1991a). The data were analysed using the steady-state assumption and droppings accumulation. On September 12th, the roads were cleared of droppings to obtain the exact period of accumulation for the second count.

Calculations required for the steady-state assumption:

1. The area sampled (*A*) is 180.1 (length) x 0.0022 (width) = 0.39622 km²,

2. The number of droppings in the whole park is (Y = 470 km² (size of park)/0.39622 km² (size of sample area)) x 54 (number of dropping counted) = 64,055

3. The number of elephants is (64,055 x ln2/49.6) / 15.0 = 59.7,

4. The variance can be approximated by: $CV^2(E) = CV^2(A) + CV^2(Y) + CV^2(r) + CV^2(D)$

   $CV^2(A) = (0.2/180.1)^2 + (0.0002/0.0022)^2 = 0.008266$
   $CV^2(Y) = (2/54)^2 = 0.001372$
   $CV^2(r) = (3/49.6)^2 = 0.003658$
   $CV^2(D) = (1/15)^2 = 0.004444$

   $CV(E) = \sqrt{0.017740} = 0.133193$

   CV(E) = SE/Mean

   SE(E) = 59.7 x 0.133193 = 8.0

   95% CL = 1.96 x 8.0 = 15.7

Result: 60 ± 8 elephants (95% CL 44 – 76), on a year-round basis there are definitely at least 44 elephants in Kasanka, possibly as many as 76, but the true population size is probably around 60.

Calculations required for droppings accumulation:

1. Area sampled (*A*) and droppings in whole park (*Y*), (*see above*),

2. With a mean daily rate of decay 0.006, and a total accumulation period (*t*) of 101 days (from the first month of the dry season, June to the beginning of the survey on 9 September 1999). The equation is as follows:

$$E = \frac{64{,}055}{(1 - 101 \times 0.006)101 \times 15} = 107.3$$

$$SE(E) = 107.3 \times 0.133193 = 14.3$$

$$95\%\ CL = 1.96 \times 14.3 = 28.0$$

Result: 107 ± 14 elephants (95% CL 79 – 135), which is somewhat higher than the estimate obtained by using the steady-state assumption.

The second dropping count was conducted from 8 to 10 October, after 28 ± 1 days of accumulation of droppings (the road was cleared on 12 September). During this period, droppings that were deposited were regularly checked for decomposition. Except for being flattened, none of the droppings showed any signs of decay and after 28 days all droppings could be identified by the observers. Calculations are the same as those done previously for droppings accumulation (*see above*) but do not include decay.

The result was 76 ± 9 elephants (95% CL 58 – 94). Because decay was zero, and the period of accumulation is exactly known, this is considered the most accurate result. Note that 2 different counts and three different methods of analysis gave almost similar results.

9.5 Sources of Error

Dropping counts are open to various sources of error. The most important one is assuming a steady-state when the system is not in equilibrium. Dropping counts in woodland savanna, conducted halfway through the dry season, will give accurate results with the steady-state method. In forest, however, the steady-state assumption should be avoided except in the second half of the dry season and dropping density should be corrected for rainfall in the preceding two months (Barnes *et al.*, 1997b). The other sources of error result from using defecation rates and decay rates from other sites. When sufficient time and money is available, site-specific rates should be measured. However, if this is not the case, the average defecation rate and the decay rate provided in this chapter can be used for dry-season counts in woodland savanna. Wet season rates should be measured on site. For forest, the average defecation rate provided in this chapter can be used, whereas the rate of decay does not need to be measured when using the rainfall equation (Barnes *et al.*, 1997b).

In addition, regardless of the technique and sample unit, the following three sources of error may occur with dropping counts (Jachmann and Bell, 1984):

1. Multiple-counting error: elephants tend to defecate while walking so that boli from one defecation are often spread over a considerable distance and may be recorded as belonging to separate defecations. As a rule of thumb, a single dropping or dung-pile consists of 5 to 6 separate boli.

2. Double-counting error: while counting, each observer should be instructed to count droppings only to one side. This error is small when using line transects, but may be large when using plots as the sampling unit.

3. Undercounting of droppings of young calves: droppings of calves that are still suckling are produced more slowly and decompose faster than large fibrous droppings, and are often missed because of their small size. Droppings counted may be increased by 5% to compensate for this error.

9.6 Faecal Counts of Other Species

When applying the dropping technique to other African animals, the same procedures and the same analysis can be used as for elephants. The only difference is the use of defecation rates and dung decay rates for the species and habitat types under study. For a few species, these rates are provided in the literature (Table 9.5).

The cost of a dropping count depends on the local labour costs and on transport costs and is more or less the same as for line-transect counts and road counts (*see Table 4.7*).

Table 9.5: Defecation rates and decay rates from the literature for several African species. Defecation rates in pellet groups and droppings (buffalo) per day, and decay rates in faecal material lost to decomposition per day.

Species	Def. Rate	Decay Rate	Habitat	Source
Kirk's Dikdik	3.4	-	Savanna	Boshe and Lyimo (1983)
Small Duikers	4.9	0.115	Forest	Koster and Hart (1988)
Large Duikers	4.4	0.110	Forest	Koster and Hart (1988)
Bushbuck	19.0	0.032*	Forest	Plumptre and Harris (1995)
Buffalo	5.1	0.021*	Forest	Plumptre and Harris (1995)

* *Exponential decay.*

Chapter 10

FOOTPRINT MEASUREMENTS

Animal spoor or footprints per unit area can be used as an index of abundance (Van Dijke *et al.*, 1986; Koster and Hart, 1988), while footprint measurements can be used to approximate the age-structure of an elephant population (Western *et al.*, 1983), or to determine absolute density and distribution of elephant and black rhino (Kelly and Beer, 1994; Jachmann, 1984a), or large cats, such as lion, leopard and cheetah (Smallwood and Fitzhugh, 1993). Both elephants and rhinos are sufficiently heavy to render footprints visible for extended periods in a variety of soil types and habitats. When the population is small (< 40 individuals) and isolated (no migration), footprint measurements provide a means to identify individual animals. This technique may be combined with dropping measurements (circumference measurements of individual boli) to estimate abundance. Although the technique is relatively simple, leading to accurate estimates of abundance when used for small populations of solitary black rhino, its applicability for a gregarious species such as elephants is more complicated. Due to its limitations, the technique does not have a wide application in the field, but nevertheless may be useful under certain conditions. With only few small isolated pockets of black rhinos remaining in the wild, these conditions are found more frequently.

10.1 Concept, Sample Design and Analysis

Footprint measurements of fore-feet and hind-feet diameter and the difference between these (Δf), provide a means to differentiate between individuals of small isolated populations of rhinos and elephants (Jachmann, 1984a). The same applies to dropping measurements, using the circumference of individual boli of the same dung-pile (Jachmann and Bell, 1984). Rhino defecate in the same places, called middens, and tend to scatter their droppings with their hind-feet. However, they often miss a few boli, which can be measured for circumference. The technique will give reliable results when used for small populations with fewer than 40 individuals.

Observers walk along grid lines placed systematically at short intervals. Intervals or spacing between lines depends on the size of the study area, the precision required, and the funds available. A team of several observers, walking side by side with approximately 10 m between counters, covers each line on foot. Using a compass or a GPS, a dead straight line is maintained. Each team member searches the ground between himself and the next person on one side only. Footprint measurements, bolus circumference and co-ordinates of observations are recorded. Footprints should only be measured when the difference between fore-feet and hind-feet can be determined to provide Δf. Diameter of footprints is measured across the widest point between left and right toe. As an additional variable, the length of fore- and hind-feet can be measured. Bolus circumference is measured around the centre of the cylinder.

For populations with fewer than 10 individuals, frequency diagrams of footprint measurements and bolus circumference may be adequate to differentiate between individual animals. For larger populations, the information should be entered into a five-dimensional array with fore-foot diameter, hind-foot diameter, Δf, bolus circumference, and co-ordinate (longitude and latitude) of each observation as variables. As additional variables, fore-foot and hind-foot length may be entered. Using statistical software such as "Statistica" or "SPSS", a cluster analysis can be performed. With solitary species such as black rhino, the result is a series of five-dimensional clusters (seven dimensional including lengths), each representing an individual animal and its activity area. With a gregarious species such as elephants, each cluster represents a group of animals and the activity area. Multiplying the number of clusters by the approximate mean group size will give an estimate of elephant numbers.

An example of the procedure required for a population with fewer than 10 animals is the estimate we did of black rhino in Mwabvi Game Reserve in Malawi in 1983 (Jachmann, 1984a).

10.2 Field Example

In the 1970s, several researchers visited the small Mwabvi Game Reserve (351 km^2) and estimated between 4 and 30 rhino (Ridding, 1975; Parker, 1976). In the early 1980s, an extension of the reserve was proposed and the authorities requested an investigation into the exact number of rhino. Because rhino mainly used the thicket vegetation, conventional techniques were not expected to provide reliable results. Instead, footprint measurements and bolus circumference measurements were used (Jachmann, 1984a).

The reserve is situated in the Lower Shire Valley in southern Malawi, on the border with Mozambique. The vegetation is a mosaic of open mixed woodland, dominated by *Albizia*, *Acacia* and *Combretum* species, interrupted by stretches of riverine thicket and small stands of mopane woodland.

Using an imaginary grid, parallel lines were placed at 1 km intervals, running from north to south and from west to east. Each line was covered by a game scout, a carrier and the author, walking side by side with 10 m between counters, giving a total line width of 30 m. Rhino spoor was followed as long as footprints could be measured. When deviating from the line of travel, the departure point was marked. Footprints and droppings were measured, preferably relating a particular spoor to the circumference of intact boli found in a midden of the same rhino. Means and standard errors were calculated for fore- and hind-feet diameter, as well as Δf, the difference between the diameters of fore- and hind-feet.

By comparing the means of the different sets of footprints measured, in combination with differences in Δf, the minimum number of rhino was estimated at 5 (Table 10.1). However, it is most likely that a few young rhino were missed, because their footprints are not as clear as those of the older and heavier animals.

Table 10.1: Mean fore-foot and hind-foot diameters with standard errors and differences between fore- and hind-feet (Δf) for 5 rhino in Mwabvi Game Reserve (Jachmann, 1984a).

	Diameter (cm)		
Rhino	**Fore-feet**	**Hind-feet**	**Δf**
1	20.40 ± 0.63	18.52 ± 0.32	1.88
2	22.14 ± 0.61	19.47 ± 0.52	2.67
3	24.06 ± 0.41	21.20 ± 0.38	2.86
4	25.50 ± 0.58		
5	18.20	16.50	1.70

Circumference measurements of 57 boli show a number of distinct peaks (Figure 10.1). Using footprint measurements found near fresh droppings, it was possible to relate bolus circumference to footprint size for rhinos 1 to 4 (Table 10.1 and Figure 10.1).

For rhino 1, circumference measurements ranged from 36 to 38 cm, for rhino 2 from 40 to 43 cm, for rhino 3 from 44 to 47 cm, and for rhino 4 from 48 to 51 cm (Figure 10.1). Regression analysis of bolus circumference on footprint diameter for rhinos 1 to 4 gave a highly significant relationship ($P<0.001$, $r = 0.997$; y (fore-foot diameter) = 6.456

+ 0.384x (bolus circumference)), supporting the classification. Bolus circumference measurements of rhinos 1 to 4, each cover a range of 3 to 4 cm. The remaining boli, with circumferences ranging from 25 to 34 cm, showing two distinct peaks and one minor peak at 27, 30 and 33 cm, probably belong to two or three unidentified young rhinos. Based on bolus circumference data, we may conclude that the reserve contained 6 or 7 rhino. Combining foot-print measurements and bolus circumference measurements, we may conclude that the reserve contained between 5 and 7 rhino.

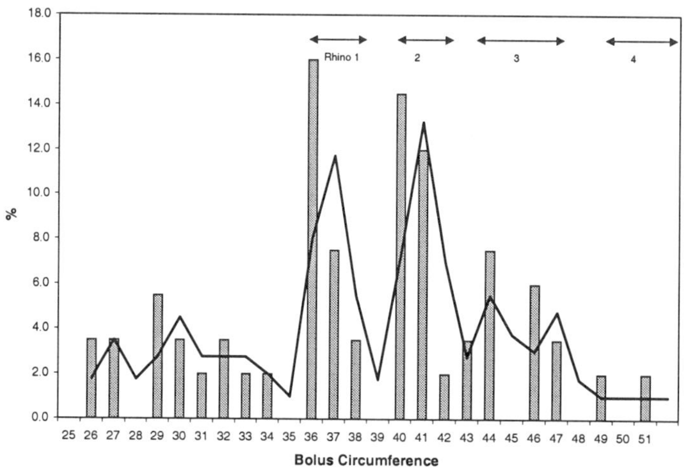

Figure 10.1: Frequency diagram of bolus circumference (57 boli) with moving average, showing the bolus circumference range for rhinos 1 to 4.

10.3 Costs of Indicator Counts

The cost of indicator counts is a function of the type of count, local salary levels, transport costs and the type of estimate required in relation to the objective. The cost of a road dropping count is the same as that of a road index count or a line-transect road count, roughly US$ 0.5 per km². The cost of a regular dropping count and a footprint survey are anywhere between US$ 0.7 and US$ 1.5 per km², depending upon the sampling intensity. Large areas will cost more per unit area than small areas, mainly because of more dead time (i.e. time lost from moving from one area to the next, and trips from and to base).

PART IV

ANALYSING DISTRIBUTION DATA AND POPULATION TRENDS

Chapter 11

ANALYSING ANIMAL DISTRIBUTION, MOVEMENTS AND POPULATION TRENDS

In previous chapters we have shown that the spatial and temporal distribution of animals has a major influence on the results of sample counts. Therefore, knowledge of seasonal movements and distribution patterns, and the environmental factors that govern these, is important to improve on survey designs.

Animals form a dynamic relationship with their environment. The basic abiotic environmental factors, such as geomorphology, soils and climate determine the types of vegetation, the availability of water and ambient temperatures, and therefore the overall pattern of spatial distribution of a species. However, animal distribution changes over time: in the long term, in response to increasing human population levels and trends in abiotic factors that influence the biotic environment, such as rainfall cycles, changes in solar radiation, erosion and tectonic movements, and in the short term, in response to fluctuations in local environmental conditions, such as seasonal differences in the availability of food and water, seasonal differences in temperature, and in response to spatial and temporal changes in illegal hunting. Spatial distribution in relation to abiotic environmental factors tends to be more or less permanent in historical terms, whereas those in response to short-term environmental fluctuations are more dynamic, and may change from year to year.

Total counts and systematic sample counts, carried out at different intervals, yield information on movements, and the spatial and temporal distribution of animals. This information can be used to improve the design of future counts, and to investigate relationships between distribution and a range of biotic and abiotic factors, and to relate distribution to management actions.

The most important abiotic factors influencing distribution of wildlife are rainfall and the availability of surface water, and nutrient and soil water availability and their effect on overall vegetation structure. Soil properties play a dominant role in plant-herbivore interactions, affecting the absolute and relative production of plant biomass and quality, which are largely determined by soil nutrient and soil water availability. Temperature is also a determinant of animal distribution through direct effects on their physiology, or by affecting resources.

Some of the biotic factors influencing animal distribution are dietary requirements, competition for food and water, reproductive strategies and predators.

Other factors with a major impact on wildlife distribution and movements are poaching, legal hunting and culling, burning, mainly through a localised and temporary reduction in food availability, and settlement. Burning can be used to manipulate vegetation structure and the temporal availability of food, and therefore animal distribution, while hunting and culling can be used to manipulate distribution by reducing animal densities in predetermined areas.

Some important reasons for studying movements and distribution are:

- *To facilitate planning of future counts.*

- *To identify unit populations.* If hunting quotas are to be set, it is necessary to know the spatial distribution of the target population. Off-take quotas should be based on a unit population rather than a variable segment of a population.

- *To determine the total range of a species.* This is important for management, such as law enforcement, planning of settlement schemes and boundaries of protected areas.

- *To identify key environmental factors underlying patterns of distribution and movements.*

- *To identify other factors influencing animal distribution and movements, such as settlement and patterns in illegal wildlife use.*

A study designed to collect information on distribution and movements requires negative observations as well as positive observations, where it is just as important to visit areas without animals as those with animals. Furthermore, cheap indirect techniques such as index counts are not suitable for collecting information on distribution and movements. Therefore, studies designed to collect distribution data require considerably more financial input than those designed for estimation of abundance. However, most total counts and systematic sample counts yield some information on distribution and movements of the target species. Therefore, this short section deals with analysis of survey results for distribution and movement patterns, and its relevance to the adaptive management system. If, however, the sole objective is to study distribution and movements, the reader is referred to Western and Grimsdell (1979) and Caughley and Sinclair (1994).

11.1 Analysis of Animal Distribution and Movements

GPS readings will give accurate locations of animal sightings during an aerial total count or a systematic aerial sample count. However, when the density of the target species is high, there is often not enough time to record GPS readings for each sighting. Then, the transect co-ordinates (start and finish), the time it took to cover the transect, and the exact time of each sighting can be used to estimate locations. For ground counts, either GPS readings or the distance covered on the transect can be used. These locations can then be used to display distribution. The most common form of displaying distribution data is a cellular pattern within a grid system design, whereby the size of the grid depends on the intensity of the count, or the interval between parallel transects. For instance, when transects are spaced at 10 km intervals, a grid size of 5 x 5 km^2 is too small, because half of the grid will lack observations. Here, the minimum grid size should be 10 x 10 km^2.

Analysis is straightforward, by enumerating all sightings for each grid square, displaying density in numbers per grid square. Analysis can be done by hand, or by using a Geographical Information System (GIS), such as ARCINFO or ARCVIEW. The advantage of computerised analysis is that distribution can be related to spatial factors by GIS overlay, and values for areas that were not sampled can be extrapolated with GIS interpolations. Although this can also be done by hand, it is much easier and more accurate with GIS.

The overall distribution of a species is mapped by averaging the densities in each grid square from a series of counts carried out at different seasons. Seasonal distribution is found by averaging densities from counts carried out in a particular season only, whereas the extent and rate of movements can be studied by comparing distribution maps made at different time intervals or different seasons.

Additional information that can be obtained from distribution data include the total density and biomass, obtained by summing the densities for each species in each grid square, and multiplying these by the mean body mass for each species. Variables that are correlated with seasonal distribution patterns may indicate causative relationships.

Most habitats are heterogeneous, with animals restricted to particular areas at particular times of the year. As a result, animal distribution is often highly uneven. There are situations, however, where it is not clear whether animals are distributed at random, or if distribution is uneven. Here, a test for randomness may be performed.

Let us assume that a population of animals is scattered over an area that is divided into hundreds of quadrats at random. The probability of settling in any one quadrat is the same for each animal and each quadrat. The frequency distribution setting out the number of quadrats with 0, 1, 2, 3, ... etc. animals on them is approximated by the Poisson series (Andrewartha, 1961). Comparison of the observed distribution with the theoretical one will show whether the target population is distributed at random or not. The formulae required to calculate the Poisson series, and those for comparing the observed and theoretical distributions, are provided in Box 11.1.

Box 11.1: Formulae for calculating Poisson series, and for comparing distributions (Andrewartha, 1961).

Poisson series:

The chance of obtaining exactly x individuals in a square is given by:

$$N\left\{\frac{e^{-m} m^{x}}{x!}\right\} \text{ where:}$$

m is the mean number of animals per square ($\Sigma x/N$),
N is the total number of squares.

With the Poisson series, the variance equals the mean, so that the ratio of the variance and the mean:

$$\frac{\Sigma(x - m)^2}{(n - 1)} \times \frac{1}{m} = 1$$

Values of less than 1 indicate that the observed distribution is more uniform, and values greater than 1 indicate that the observed distribution is more patchy than the Poisson series.

The observed distribution can be compared with the Poisson series with the Chi-squared test, whereby X^2 is estimated by:

$$\frac{\Sigma(x - m)^2}{m}$$, with (n – 1) degrees of freedom.

A large value for X^2 indicates a significant departure from the Poisson series.

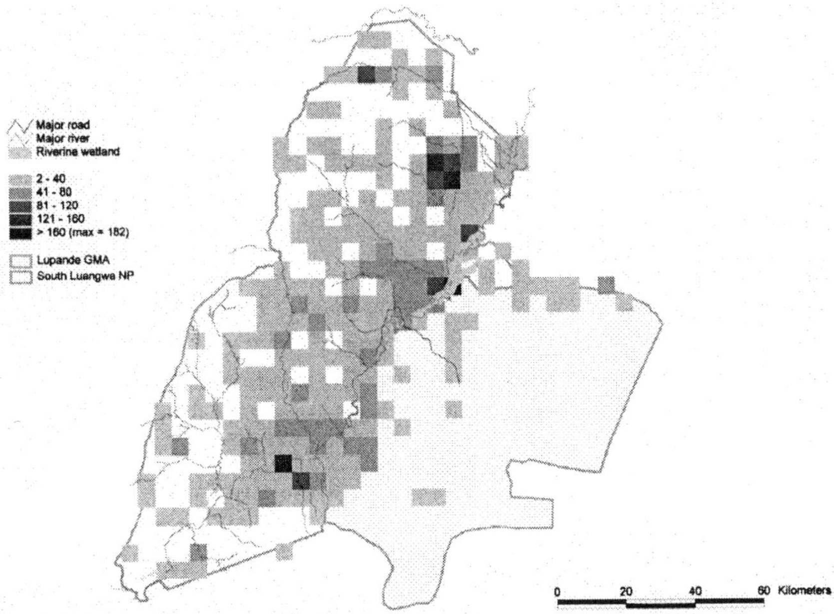

Figure 11.1: Dry-season elephant distribution in South Luangwa National Park and Lupande Game Management Area, from a series of 7 aerial surveys carried out between 1993 and 1999 (Jachmann and Phiri, 1999b). Density is in numbers per 25 km².

11.2 Field Example

As an example, we use the dry-season distribution of elephants in South Luangwa National Park and Lupande GMA (SLAMU area), obtained through a series of 7 aerial sample counts, conducted between 1993 and 1999 (Jachmann and Phiri, 1999b). Counts conducted prior to 1993 used a different design, and were not suitable for this type of analysis. Parallel transects were spaced at 5 km intervals in 1993, and at 2 minute intervals in the following years (*see Figure 5.4*). Pooling the data from a series of counts gives a more accurate account of dry-season distribution. Elephant distribution was computed with ARCINFO, using a 5 x 5 km² grid, showing the average numbers of elephants per year per 25 km² (Figure 11.1). Dry-season distribution, combined with wet-season observations (Jachmann, unpublished), will provide range use by elephants on a year-round basis (Figure 11.2).

Within the SLAMU operational area, the *Core Range*, or the area occupied by elephants throughout the year, covers most of the park and the western, alluvial part of the GMA (Figure 11.2). The *Seasonal Range*, or the area where some elephants spend part of the wet season, when water and food are abundant and tall grass provides protection against poachers, is covered by the central part of the GMA (Figure 11.2). The eastern part of the GMA forms the *Occasional Range*, where some individuals, mainly bulls, occur occasionally (Jachmann, unpublished).

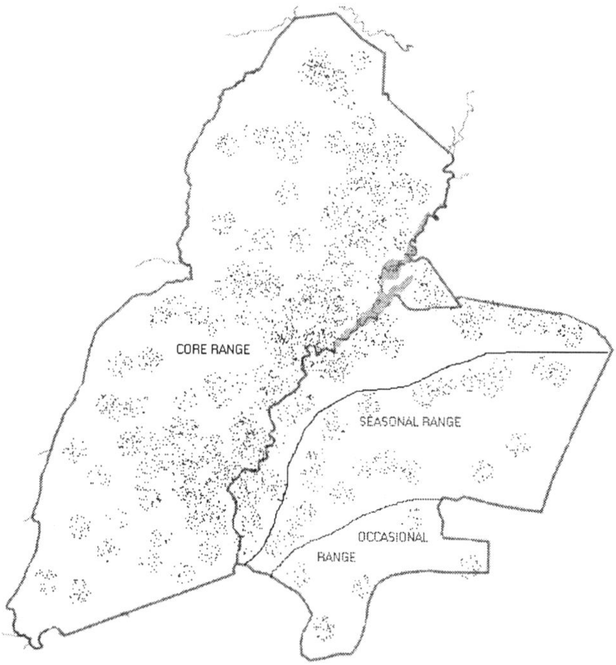

Figure 11.2: Elephant range use in South Luangwa National Park and Lupande Game Management Area, indicating Core Range, Seasonal Range and Occasional Range.

For management purposes, it is important to know whether elephants are distributed over the area at random, or if distribution is uneven. An uneven distribution may be in response to various environmental factors, such as the availability of surface water, the quantity and quality of forage, or in response to disturbance factors, such as poaching, settlement and burning. In our example we do not require statistics to show that during the dry season elephants are not distributed at random, but concentrated near the Luangwa River, with a declining density gradient towards the west and towards the east of the project area (Figure 11.1). Also, throughout most of the dry season, a large part of the Lupande GMA is not utilised by elephants (Figure 11.1).

We continue by investigating why elephants in the SLAMU area are unevenly distributed (Figure 11.1). The first spatial factors to be examined are the availability of surface water and soil nutrient status, because together they govern vegetation type and forage quality. At this stage, it is useful to look at a cross-section of the central Luangwa Valley. Each of its geomorphological regions has a different soil nutrient status and soil water availability, and therefore different habitat types (Figure 11.3).

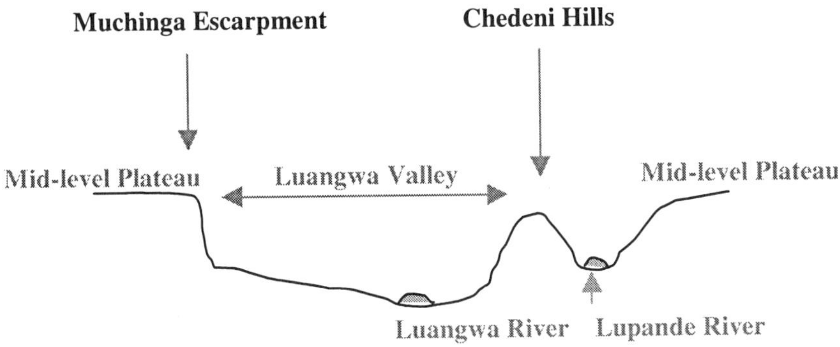

Figure 11.3: Diagram of broad geomorphological regions in the central Luangwa Valley.

The Luangwa Valley is part of the Central African Rift, where soils of recent volcanic origin, combined with alluvial soils, result in a fertile valley trough, with rich grassy plains, riverine, mopane, and mixed *Acacia* woodland (Jachmann, 2000). Towards the west, soils become gradually poorer, with Munga woodland (mixed *Combretum/Terminalia*) in the centre of the park, miombo woodland (*Julbernardia/Brachystegia*) at the bottom of the Muchinga Escarpment, and ecotone vegetation in-between. Towards the east, the Chedeni Hills are covered by poor quality shrub, with mixed mopane woodland and thickets in the narrow Lupande Valley, with miombo woodland in the foothills and on the mid-level plateau (Figure 11.3).

During the peak dry season, surface water is found in the Luangwa River and some of its main tributaries, in most streams and rivers originating on the Muchinga Escarpment, and in many natural waterholes scattered throughout the area. Thus, elephant distribution is probably not influenced by surface water. However, the first important factor influencing elephant distribution is high quality vegetation in the valley trough, within 10 km of the Luangwa River (Figure 11.4). During most of the year, the alluvial valley trough has high concentrations of elephants. This, however, does not explain why there are many areas with ample food and water, but without elephants (Figure 11.1). Some of these areas, such as in the Lupande GMA, have large settlements with a high human density, which explains why elephants only utilise them in the wet season, when crops are abundant, and tall grass provides protection (Figure 11.4). Other areas, however, lack settlement and human presence, but still contain much lower densities of elephants than one might expect. These are mostly areas that are not frequently patrolled by the scout force, with a continuous influx of poachers from towns located on top of the Muchinga Escarpment in the west of the park, or from the settlements in the Lupande GMA (Figure 11.4). Thus, the third important factor influencing elephant distribution is disturbance by illegal activity (Jachmann, 1998).

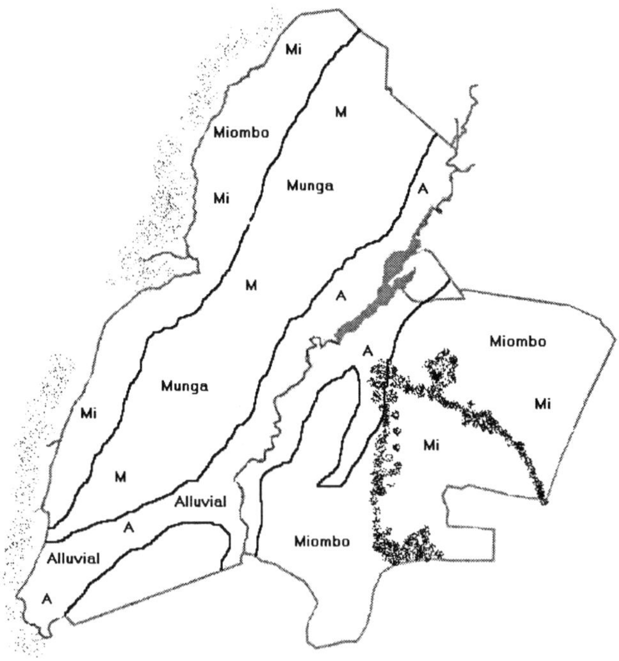

Figure 11.4: Generalised vegetation types, settlement (black dots) and main areas wit influx of poachers from outside the project area (grey dots) in South Luangwa Nation: Park and Lupande GMA.

These three factors combined - the availability of food, settlement and poaching - result in the following seasonal movements and distribution patterns. During the early wet season, from November to January, large concentrations of elephants are found in the floodplains near the main rivers, where good quality grass is abundant. During the mid and late wet season, from January to April, when grass quality sharply drops and crops are ripening, elephants gradually turn to browse and raiding of gardens, thereby gradually spreading over the project area, with some males moving towards the periphery. During this period, tall grass provides protection against poachers. During the early and mid-dry seasons, from April to August, some elephants move outside the project area, in search of suitable browse. During the late dry season, from August to November, fires have reduced the food supply in many areas, especially those with settlement and high poaching levels in the GMA and the peripheral areas of the park. Fires open up the vegetation by removing the grass layer and part of the foliage, which makes elephants more vulnerable to poaching. During this period, elephants will gradually move back to the relative safety of the park and the eastern part of the Lupande GMA, where food is also more abundant due to fire protection.

This information is important for adaptive management for several reasons: first it shows that the peripheral areas of the park and most of the GMA require more patrols, intensifying the law-enforcement effort, and second it shows that future aerial survey designs should include stratification.

11.3 Analysing Population Trends

Most wildlife counts are, or should be, part of a monitoring programme, established to detect trends in population numbers. Whatever the objective for the area, it is always important for management to know whether a particular population is declining, increasing or remaining stable. Different techniques and varying efforts give results with different levels of accuracy and precision, making it difficult to compare estimates. Trend analysis for a particular species requires the use of a single technique and standardised procedures, resulting in estimates with similar levels of bias and precision. When counting large migratory animals such as elephants, and when the survey area does not cover the total range, the most important factor is timing of the survey. Animal distribution changes constantly in response to seasonal fluctuations in environmental conditions. When good quality grass is abundant, distribution may be clumped. When grass quality deteriorates, elephants turn to browse and spread over a large area, mainly due to food competition. Some animals may leave the survey area altogether, resulting in underestimates, while a clumped distribution will result in estimates with a low precision. Timing of surveys is a lesser consideration when counting smaller sedentary species, or when the survey area includes the entire range of the target species.

A population may be declining, stable, or increasing. For our purposes, there are three relevant measures of increase (Caughley, 1977). *The finite rate of increase* (λ) is the ratio of numbers in two successive years. This is better expressed as the exponential rate of increase (r) relating numbers at two points in time. *The intrinsic rate of increase* (r_m) is the exponential rate at which a population with a stable age distribution grows when no resource is in short supply. The *observed rate of increase* ($\check{r}$) is the exponential rate at which a population increases over a period of time. It does not imply that it is constant over this period, that the age distribution is constant or that resources are in ample supply. Although the two exponential rates r_m and $\check{r}$ are mutually dependent and sometimes the same, we assume that analysis of population trends, based on survey data, results in the observed rate of increase.

With only two estimates obtained in two successive years, the rate of change is given by the exponential rate of increase (r):

$$r = (\ln N_t - \ln N_0)/t \qquad \text{where:}$$

N is numbers, and
t is time.

If we wish to test whether the rate of change or the difference between two estimates is significant, assuming a normal distribution, we will first have to test for equality of variances. Testing is required to examine whether any differences between population estimates are due to sampling error alone, or to real differences, such as migration for instance. This is done by the F-test, which is the ratio of the larger to the smaller variance:

$$F = Var_1/Var_2 \qquad (Var_1 > Var_2)$$

Degrees of freedom: numerator $f_1 = n_1 - 1$
denominator $f_1 = n_2 - 1$

If the tabulated value for the chosen significance level is lower than the observed value (F), the two variances are significantly different. In this case we treat this as being distributed approximately like 'student's' t with f degrees of freedom (*see Box 11.2*). If, however, the two variances may be assumed equal, then the rate of change is significantly different at the 5% level when the ratio of the difference between the two estimates to its standard error is more than 1.96 standard errors from the mean. Graphically we can explain this as follows: if we construct two normal distributions, using the estimates and the standard errors, the curves should not overlap more than at the 5% level (1.96 SE) to be significantly different (two-tailed). For example, two imprecise estimates, one with mean 100 and standard error 40, and the other with mean 120 and standard error 40, each with 30 observations, have normal curves of which the 5% tails overlap (Figure 11.5). These estimates are not significantly different. Increasing the precision by reducing the standard errors for both estimates to 20 (Figure 11.6) gives a significant difference ($P < 0.001$). The formula for comparing two estimates with known variances that are assumed equal is shown in Box 11.2.

Box 11.2: Formulae for comparing two estimates with known variances (Bailey, 1981).

Assuming that the estimates x_1 and x_2 are each normally distributed, the ratio of these estimates (d) is also a normal variable with a true mean of zero and a unit standard error:

$$d = \frac{x_1 - x_2}{\sqrt{(Var_1^2/n_1) + (Var_2^2/n_2)}}$$

If the absolute value of d is greater than 1.96, then there is a significant difference between the two estimates. The 95% confidence limits of the difference between the two estimates ($x_1 - x_2$) can be estimated by multiplying the denominator of the above equation by d.

With a significant difference in variances, d is treated as being distributed approximately like 'student's' t, with f degrees of freedom, where:

$$f = \frac{1}{\dfrac{u^2}{n_1 - 1} + \dfrac{(1-u)^2}{n_2 - 1}} \quad \text{and} \quad u = \frac{Var_1^2 / n_1}{Var_1^2 / n_1 + Var_2^2 / n_2}$$

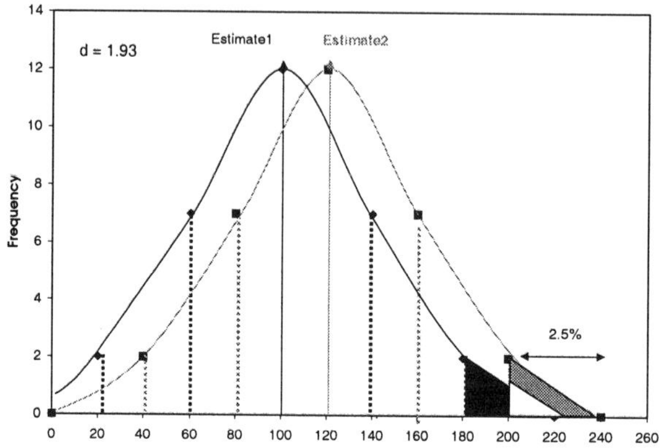

Figure 11.5: The normal curves for two estimates with means 100 and 120, and standard errors 40 overlap at the 5% level (two-tailed). The estimates are not significantly different.

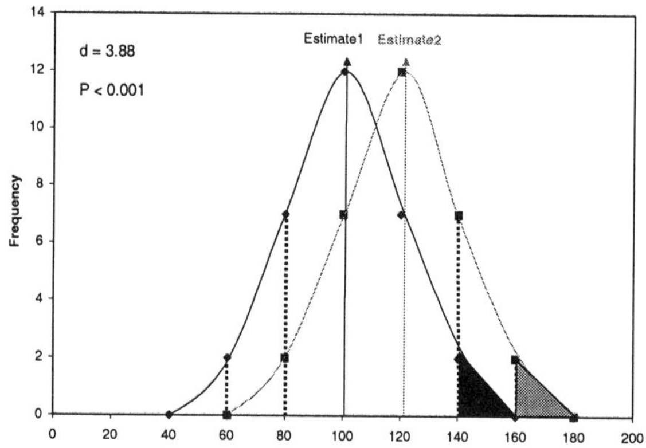

Figure 11.6: The normal curves for two estimates with means 100 and 120, and standard errors 20 do not overlap at the 5% level (two-tailed). The estimates are significantly different ($P < 0.001$).

With more than two estimates, and in the absence of more complex patterns of change, the easiest way to calculate the observed rate of change is exponential regression, or linear regression of the natural logarithms of the estimates versus time. This presents population change in terms of percentage change per unit time. The observed rate of change can also be calculated with the formula provided in Box 11.3.

Box 11.3: Formula for calculating the observed rate of change of a population (Caughley, 1977).

Estimates are converted to natural logarithms (ln), each designated N (n estimates).
Time units of one year, where the first equals 1, will each be designated t.

The mean exponential rate of change is:

$$\check{r} = \frac{\Sigma N_t - \{(\Sigma N)(\Sigma t)/n\}}{\Sigma t^2 - \{\Sigma t)^2/n\}}$$

When the survey area covers only part of the total range of a migratory species, minor differences in sampling intensity and precision among estimates are lesser considerations than timing of surveys. For example, analysis of elephant population trends over its total range in Botswana used 14 estimates obtained over a period of 9 years, with the estimates differing in sampling intensity and precision (Gibson and Craig, 1998). To check the validity of using exponential regression, a Monte Carlo simulation was conducted. For every result, a normal distribution was constructed using the estimate and its standard error. From each of these distributions, a possible estimate was drawn at random, and a regression coefficient calculated. This was repeated a million times, resulting in a mean regression coefficient and its variance.

The estimated rate of increase was 6.05% (range 3.1 – 9.1%), which compares with 6.00% (range 2.9 – 9.2%) for the exponential regression (Gibson and Craig, 1998).

11.4 Field Example

To illustrate potential pitfalls in trend analysis, we will use the results of a series of 9 aerial sample counts of elephants, conducted in South Luangwa National Park and Lupande GMA (SLAMU area) between 1989 and 1999 (Table 11.1). Starting with the estimate obtained in 1989, predicted numbers were simulated with a 5% rate of increase (Table 11.1). During this period, the numbers of elephants known to have been killed by poachers was always less than 0.25% per year (Jachmann, 1998 and unpublished).

Table 11.1: Results of 9 aerial sample counts of elephants in South Luangwa National Park and Lupande GMA, conducted between 1989 and 1999.

Year	*Month*	*Population Estimate*	*Standard Error*	*%SE*	*Predicted Numbers for 1999*
1989	February	5,408	790	14.61	5,408
1990	October	5,500	878	15.97	5,678
1991	January	5,566	1,382	24.82	5,962
1992	-	-	-	-	6,260
1993	June	5,929	1,032	17.41	6,573
1994	October	8,026	1,240	15.45	6,902
1995	-	-	-	-	7,247
1996	October	8,834	1,655	18.73	7,609
1997	September	8,499	1,951	22.96	7,990
1998	October	7,370	912	12.37	8,389
1999	July	6,301	533	8.46	8,808

From 1989 to 1991, the surveys were conducted by a different team, using a different design, than those carried out after 1991. On average, sampling intensity was between 5% and 10%, calibrated strip width between 140 m and 190 m, and height between 100m and 150 m above ground level. The double-count technique was used for all surveys. In the Luangwa Valley, total elephant range covers approximately 80,000 km². The survey area was approximately 15,000 km². Timing of surveys ranged from mid-wet season (January) to late dry season (October).

Thus, the data-set is far from perfect, with different survey designs and considerable differences in timing each year, only covering 19% of the total elephant range. Exponential regression of elephant numbers versus time is not significant, and the relationship is best described by a polynomial curve (Figure 11.7). Some estimates deviate considerably from the number predicted for that year assuming a 5% annual increase (Figure 11.7).

As shown in the previous chapter, the distribution of elephants in the central Luangwa Valley changes from one season to the next. Therefore, let us investigate the influence of timing on the survey results.

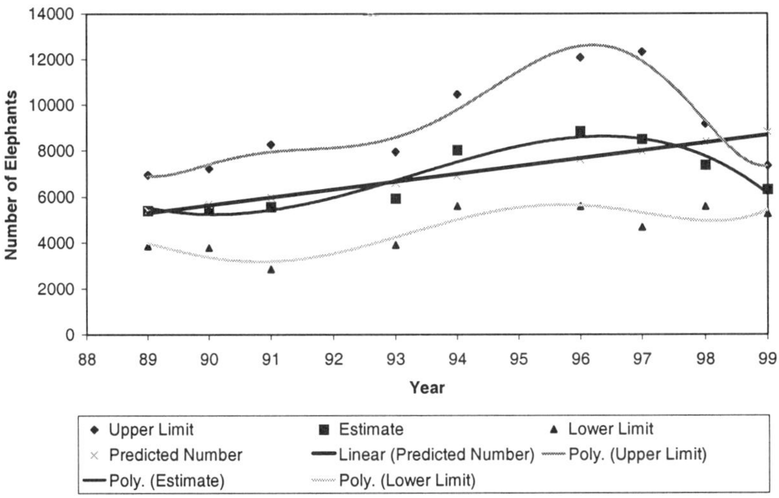

Figure 11.7: Elephant population estimates for the SLAMU area, obtained by aerial survey from 1989 to 1999. A polynomial has been fitted to the estimates. The straight line shows the trend in elephant numbers expected if the population increased by 5% each year.

With a simple simulation, using a 5% rate of increase, each of the 9 population estimates (Table 11.1) was extrapolated to give the predicted number of elephants in 1999, for the month the survey was initially carried out. There were four surveys conducted in October, so the 1999 estimate for October was the average of those predictions. This results in predicted elephant numbers in 1999 for 6 different months of the year (Figure 11.8). This procedure gives a rough idea of the influence of timing of the survey on the numbers of elephants estimated.

On average, surveys carried out during the late dry season (September – October) give the highest predictions, surveys conducted during the early and mid-dry season (April – August) give the lowest predictions, and those done in the mid-wet season (January – February) give intermediate predictions (Figure 11.8).

This closely follows the seasonal changes in elephant distribution described in the previous chapter, with a proportion of the population utilising peripheral areas outside the park during the early and mid-dry season, resulting in low estimates, and most of the population concentrated in the eastern part of the park and the western part of the GMA in the late dry season, resulting in high estimates (Figure 11.8).

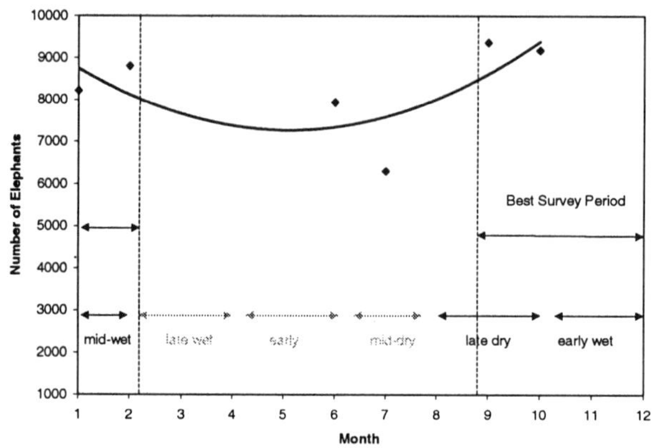

Figure 11.8: The influence of timing of aerial surveys on the size of the population estimate for elephants in the SLAMU area.

If we omit the results of the two surveys conducted in June (1993) and July (1999), because they were done at the wrong time, a highly significant ($P < 0.001$) exponential relationship emerges, explaining about 77% of the variability in the data ($R^2 = 0.7669$, Figure 11.9). Regression analysis suggests that between 1989 and 1998 the elephant population in the SLAMU area has increased by about 5.3% per year (Figure 11.9).

This illustrates the importance of standardised procedures and designs when examining population trends. When the study area covers only part of the range of a migratory target species, timing of the count is probably the most important consideration.

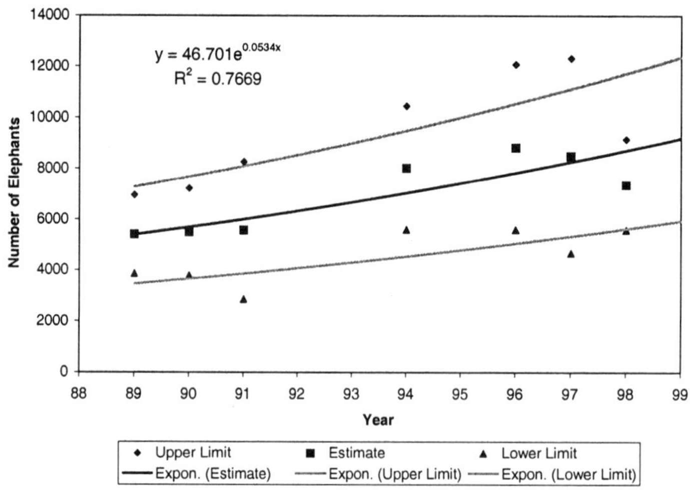

Figure 11.9: Exponential regression analysis of elephant numbers versus time, for 7 aerial sample counts conducted in the SLAMU area between 1989 and 1998.

PART V

SYNTHESIS

Chapter 12

DECIDING ON A TECHNIQUE

This subject was briefly introduced in the first chapter (*see section 1.2*), but now needs to be expanded to relate the different techniques for estimating animal abundance to conditions commonly found in practice. When deciding on a technique, there are 4 important points to be considered:

1. For what reason do we require an estimate of abundance? This is the single most important consideration, because it determines the type of estimate required in terms of precision and accuracy. Depending upon the species and the habitat, different techniques produce results with different levels of precision and accuracy. Thus, the objective of the count is the first consideration that will restrict our options in terms of suitable techniques.

2. The animal species that needs to be counted, its density and distribution over the survey area, will further limit the scope of techniques that will produce an estimate that satisfies the objective of the count.

3. The type of habitat, in terms of topography and vegetation density, will impose further restrictions on the scope of methods that can be applied.

4. Further limitations may be imposed by the budget and logistics. After careful consideration of the first 3 points, the conclusion may be a technique that is too expensive. A cheaper technique may not give adequate results. In that case, it is better to cancel the exercise, rather than spend limited resources on collecting information that does not satisfy the objective.

Each of these points will be discussed briefly. A series of diagrams showing the sequence of decisions by which a technique is chosen will aid selection of a technique for a particular set of conditions (Figures 12.1 to 12.8).

12.1 The Reasons for Counting

Because the reason for counting determines the type of data required, it also determines the technique and the level of effort required. Some of the common reasons for counting animals, the type of information required, and the appropriate techniques are listed below:

- *Regulation of wildlife-based industries*: Here, the goal is to set quotas for hunting off-take or population reduction by culling. For this purpose we require annual estimates for all relevant species, with a high level of accuracy. Many techniques generate estimates that are not very accurate, and tend to underestimate the population in question. Poor estimates rarely lead to ecological disaster, but more commonly to conservative off-take quotas. Depending upon the species and habitat, line-transect methodology, aerial techniques and indicator techniques, such as dropping counts, can be used. However, if carried out in a consistent manner, trophy quality and hunting success indices will provide management with information that is both cost-efficient and of an adequate quality.

- *Regulation of management practises*: Here, the goal is to monitor the effect on levels of illegal off-take of changes in law-enforcement patterns, such as patrolling effort and efficiency, the impact of community programs, and international legislation, such as the CITES ivory ban. Estimates should be precise, but not necessarily accurate, and obtained with a high level of consistency in methodology applied, using large samples. Estimates of animal abundance for this purpose should ideally be collected through cost-efficient index techniques, for example patrol index, where sightings of key species are related to patrol effort.

- *Conservation concerns*: Often we need to know whether the abundance of a species of particular interest is below or above its 'limits to acceptable change'. Estimates need neither be precise, nor accurate, and therefore index techniques usually provide adequate information. A problem arises when there is a sense of urgency, and no previous monitoring has been done. Depending upon the species and the habitat, low-intensity aerial survey techniques may then be the quickest and cheapest solution.

In summary, not all reasons for counting require accurate estimates of population size. Frequently, an index of trends that can be detected with confidence over a five-year period is sufficient.

12.2 The Animal Species

Different methods are needed for each species. The appropriate method for a particular species depends on a number of behavioural factors, such as the time of day the species is active (*nocturnal* versus *diurnal*), whether it is *sedentary* or *migratory*, whether it is *solitary* or lives in *groups*, the average number of animals in a group, the *conspicuousness* of the species, i.e. the body size and colour patterns of the animals, its spatial distribution, and its density. Some of these factors are inter-related, e.g. nocturnal species are often solitary, and solitary species are usually sedentary. These behavioural attributes are related to physical factors, such as the diet, body size, reproductive and protective strategies, but also to environmental factors, such the availability of food and surface water, and disturbance factors, such as poaching. To illustrate this, 6 out of 9 large carnivores occurring in Zambia are nocturnal sedentary animals, predominantly leading a solitary life (Table 12.1). The three other carnivore species live in medium-sized groups; e.g. lion live in prides numbering up to roughly 30 individuals, cheetah usually live in pairs, but sometimes in small groups up to 6 individuals, while wild dog live in packs numbering up to about 40 individuals. Because of their body size, mean group size and behavioural characteristics, carnivore species need to be counted with indirect techniques, or by using individual recognition.

Indirect techniques can be applied with most large mammal species. However, these techniques have a number of disadvantages. For example, the capture/recapture technique only works under specific conditions that are difficult to satisfy in the field. Faecal count techniques, foot print measurements, radio telemetry and DNA analysis are very labour intensive, and often too costly to deploy, whereas index counts require consistency in methodology, timing of counts and other related parameters. However, index techniques are much more cost efficient than direct techniques, and when used in a consistent manner, they often provide adequate information on trends. Therefore, you are strongly advised to use index methods whenever possible.

High sampling errors and the risk of large biases mean that aerial counts are not recommended for most species. Also, aerial techniques are expensive, and often unnecessary when referring to species over which there is no need for specific management procedures, such as quota setting. In addition, including these species in an aerial survey programme is counterproductive, since it reduces the accuracy of key species estimates.

Because this manual has a heavy concentration of examples from Zambia, we will now take a closer look at selection of techniques appropriate for particular wildlife species occurring in Zambia. Most of these species also occur elsewhere on the continent.

Table 12.1 provides a summary of large mammal species found in Zambia, the average weight range for adult males and females combined, the average group size category, where S is solitary, S_m is small (<10), M is medium (10 – 30), and L is large (>30), whether Direct (D) or Indirect (I) methods should be used, and the appropriate technique(s).

As mentioned above, carnivores need to be counted with indirect techniques, such as index and radio telemetry, or with individual recognition. Under circumstances when index techniques cannot be used, assessing abundance of carnivores with the other two techniques is an expensive undertaking.

Elephants can be counted with most techniques; the appropriate choice depends on the objective and the habitat type. In the rainforest, however, the only options are dropping counts, or individual recognition for small populations.

The abundance of black rhino can only be assessed with indirect methods, such as index techniques, mark/recapture, radio telemetry, DNA analysis, faecal and footprint methods for small populations, and individual recognition, sometimes combined with mark/recapture. The abundance of white rhino, however, can be estimated with direct techniques, such as line-transect methodology.

Zebras can be counted with most direct techniques, while individual recognition may be appropriate under conditions of low density or thick vegetation.

The abundance of lechwe can be estimated with most direct techniques. However, with aerial counts, photography should be used to accurately count the individuals in large herds.

The abundance of small, mostly solitary and sedentary antelope species, such as duiker, bushbuck, klipspringer, grysbok and steenbok, and also warthog, can be assessed with line-transect methodology or faecal methods. Bushpig is a special case, where abundance can often only be assessed with particular indirect techniques, such as radio telemetry, but in some cases line transects may work.

Some of the larger herbivores, such as eland, wildebeest, waterbuck, hartebeest, roan, sable and tsessebe can be counted with most direct techniques. Line-transect methodology should be deployed in dense vegetation. However, whenever possible, cost-efficient index techniques should be deployed, such as patrol or road count indices. Estimation of kudu abundance requires index techniques, individual recognition or line-transect methodology. Although line-transect methodology can be used for giraffe, individual recognition may give better results.

Table 12.1: Large mammals occurring in Zambian conservation areas, the average weight for adult males and females combined, average group size (**S** is Solitary, **Sm** is <10, **M** is 10 – 30, **L** is >30) and guidelines as to the techniques appropriate for estimating their abundance (**D** is Direct and **I** is Indirect Technique). **All** is all direct techniques can be used, **AP** is aerial sample or total count, using photography, **FF** is faecal and footprint techniques, **IR** is individual recognition, **L** is line-transect methodology, **RT** is radio telemetry, including GPS tracking, **SG** is sample ground count, **TG** is total ground count, and **X** is index techniques.

Species	Average Weight (kg)	Group Size	D/I	Appropriate Techniques
Carnivores				
Aardwolf *(Proteles cristatus)*	11-14	S	I	X,RT,(IR)
Caracal *(Felis caracal)*	16-18	S	I	X,RT,(IR)
Cheetah *(Acinonyx jubatus)*	45-64	S/Sm	I	X,RT,(IR)
Leopard *(Panthera pardus)*	50-82	S	I	X,RT,(IR)
Lion *(Panthera leo)*	123-204	Sm	I	X,RT,(IR)
Serval *(Felis serval)*	14-18	S	I	X,RT,(IR)
Side-striped jackal *(Canis serval)*	9-10	S/Sm	I	X,RT,(IR)
Spotted hyena *(Crocuta crocuta)*	45-80	Sm	I	X,RT,(IR)
Wild dog *(Lycaon pictus)*	25-32	Sm/M	I	X,RT,(IR)
Proboscidae				
Elephant *(Loxodonta africana)*	1,500-3,000	M	I/D	All,IR,RT,FF
Odd-Toed Ungulates				
Black rhino *(Diceros bicornis)*	900-1,400	S	I/D	L,X,RT,FF
Burchell's zebra *(Equus burchelli)*	225-320	M	D	All
Even-Toed Ungulates				
Black lechwe *(Kobus leche smithemani)*	77-120	L	D	All,AP
Blue duiker *(Philantomba monticola)*	4-9	S	D	L,FF
Blue wildebeest *(Connochaetes taurinus)*	160-272	L	D	All
Buffalo *(Sincerus caffer)*	800-1,000	L	D	All,AP
Bushbuck *(Tragelaphus scriptus)*	32-77	S	D	L
Bushpig *(Potamochoerus porcus)*	55-82	M	I/D	X,RT,L
Cape Eland *(Taurotragus oryx)*	590-680	L	D	All
Cookson's wildebeest *(C.t. cooksoni)*	150-265	M	D	All
Common duiker *(Silvicapra grimmia)*	10-14	S	D	L,FF
Common waterbuck *(Kobus ellipsiprymnus)*	160-210	M	D	All
Defassa waterbuck *(Kobus defassa)*	160-272	M	D	All
Giraffe *(Giraffa camelopardalis infumata)*	900-1,200	M	D	L,X,IR,FF
Hippopotamus *(Hippopotamus amphibius)*	1,100-1,800	L	D	TG,SG,AP
Impala *(Aepyceros melampus)*	45-82	L	D	All
Kafue lechwe *(Kobus leche kafuensis)*	77-120	L	D	All,AP
Klipspringer *(Oreotragus oreotragus)*	14-18	Sm	D	L
Kudu *(Tragelaphus strepciseros)*	180-320	Sm	D	L,X,IR
L. hartebeest *(Sigmoceros lichtensteinii)*	120-145	Sm	D	All

Oribi *(Ourebia ourebi)*	9-20	Sm	D	L,TG
Puku *(Kobus vardonii)*	75-91	Sm/M	D	TG
Red lechwe *(Kobus leche leche)*	77-127	L	D	All,AP
Reedbuck *(Redunca arudinum)*	50-77	S	D	TG
Roan *(Hippotragus equinus)*	227-272	M	D	All
Sable *(Hippotragus niger)*	180-250	M	D	All
Sharpe's grysbok *(Raphicerus sharpei)*	7-11	S	D	L,FF
Sitatunga *(Tragelaphus spekei)*	45-110	S	D	X,RT,AP
Steenbok *(Raphicerus campestris)*	9-14	S	D	L
Thornicroft's giraffe *(G.c. thornicrofti)*	900-1,200	Sm	D	L,X,IR,FF
Warthog *(Phacochoerus aethiopicus)*	65-135	Sm	D	L
Tsessebe *(Damaliscus lunatus)*	135-160	Sm	D	All
Yellow-b. duiker *(Cephalophus sylvicultor)*	45-64	S	D	L,FF

Hippos and crocodiles need to be counted on the ground with specialised techniques, or aerial counts, using photography (in the case of hippos). Oribi, puku and reedbuck always follow an uneven distribution, with large concentrations in dambo areas and grassy plains. These species should always be total-counted from the ground. Estimation of sitatunga abundance requires specialised studies, such as radio telemetry, capture/recapture, or possibly aerial total counts, using low and slow-flying micro-light aircraft, or a helicopter (Table 12.1).

12.3 The Habitat Type

Only two variables are important when differentiating between habitat types, namely topography and vegetation. The first differentiation is between the savanna and rainforest biomes. Only indirect techniques can be used in the rainforest. Within the savanna biome, topography is the first factor to consider. If the landscape is flat to slightly undulating, there is ample choice of techniques. However, when topography is uneven, with steep hills and valleys, application of direct techniques is restricted to line-transect counts, correcting transect length for slope. The next factor is vegetation density. In areas with dense vegetation, such as for example in closed canopy miombo woodland, dwarf mopane or *Combretum* thickets, line-transect methodology or indirect techniques should be applied. In conservation areas with open woodland, some animals can be counted from the air, while others can be counted with direct ground counts using line-transects, or by indirect techniques.

12.4 The Budget

The price tag varies with the technique, and for each technique, cost is a function of the type of estimate required in relation to the objective. The type of estimate determines the effort or area coverage needed, which in turn is a function of density and distribution of the target species. In sample counts, the area coverage should ideally be between 5% and 15%. Sampling intensities higher than 15% only give minor improvements in precision at greater cost. Total counts may be done at different sampling intensities. Also, there are substantial differences in salary levels and transport costs throughout the continent. Therefore, an indication of cost range is provided for each technique, rather than a fixed amount per unit area (Table 12.2). For most of the indirect techniques, however, it is immensely difficult to provide cost per unit area, because approaches, time, labour and equipment required vary widely for different species, habitats and objectives. Therefore, for indirect techniques a relative cost indication is given, ranging from very low to very high (Table 12.2).

The line-transect foot count is one of the most expensive direct techniques (Table 12.2). However, cost indications provided in this manual are based on wildlife densities in the Nazinga Game Ranch, Burkina Faso, and the Lupande GMA, Zambia. Although we feel that these are representative of densities found elsewhere on the continent, higher densities will require fewer repeat counts to give similar levels of precision, thereby reducing cost. The major advantage of a line-transect count is the relative ease of its implementation in the field, under widely varying conditions. In areas with an extensive road-system, however, the preferred choice is a line-transect road count, significantly reducing cost (Table 12.2).

The cost of aerial techniques is mainly a function of aircraft hire and fuel prices. Aerial techniques are expensive and not very cost-efficient for the poor quality of data they return for many species. However, they may be the only alternative for covering large areas without a road-system. Then, biases may be minimised by applying the double-count procedure. Indication of cost for special aerial surveys to count hippos and crocodiles are provided per km of river frontage, and should therefore not be compared with cost per unit area as provided for the other techniques (Table 12.2).

Dropping counts are the preferred indirect technique, resulting in high quality data at the same cost as aerial techniques (Table 12.2). Provided that information is available on defecation and decay rates, dropping counts result in estimates of population size and distribution based on accumulation over periods of up to several months, in contrast to other techniques, where results are based on instantaneous distribution.

However, index counts are the preferred technique under conditions where they can be applied with a high level of consistency, producing adequate information on trends. In areas where wildlife is exploited, and/or in areas with a law-enforcement component, trophy quality, hunting success and patrol index are cost-efficient techniques, requiring little to no additional resources. Other indirect techniques, such as mark/recapture, radio telemetry and DNA analysis will always be labour intensive and costly, and should only be applied when sufficient resources are available, and when the information cannot be obtained through any of the other cheaper methods.

Table 12.2: Estimated costs or relative cost indications of different techniques (ranging from very low to very high).

Technique	Cost per km^2	Cost Indication
DIRECT COUNTS		
Line Transect Foot Count	US$ 4.0 – 5.0	
Line Transect Road Count	US$ 0.5 – 1.0	
Individual Recognition		High
Aerial Sample Count	US$ 0.7 – 1.2	
Aerial Total Count	US$ 1.9 – 2.5	
Aerial Special Count*	US$ 1.0 – 1.5	
(Hippos & Crocodiles)		
INDIRECT COUNTS		
Mark/Recapture		Very High
Radio Telemetry		Very High
DNA Analysis		Very High
Patrol Index		Very Low
Road Index	US$ 0.2 – 0.5	
Trophy Quality		Low
Hunting Success		Very Low
Dropping Count	US$ 0.7 – 1.5	
Footprint Count	US$ 0.7 – 1.5	
Dropping Road Count	US$ 0.5 – 1.0	

* Cost per km of river frontage.

12.5 Making a Decision

In this last section we will provide diagrams displaying the sequence of decisions by which a technique is chosen (Figures 12.1 to 12.8). Large mammals are arbitrarily grouped into 10 categories, each of which is counted with similar techniques. A reference to the chapter and section describing the technique is shown between brackets. The categories are as follows:

1. *Carnivores* (Figure 12.1).

2. *Large Conspicuous Gregarious Animals*: Elephant, Buffalo, Wildebeest, Zebra and Lechwe (Figure 12.2).

3. *Large Gregarious Animals*: Giraffe, Eland, Hartebeest, Roan, Sable, Waterbuck, Tsessebe and to some extent Impala (Figure 12.3).

4. *Small Sedentary and Usually Solitary Animals*: Duiker, Bushbuck, Klipspringer, Steenbok, Grysbok and Warthog (Figure 12.4).

5. *Animals with a Clumped Distribution*: Puku, Reedbuck and Oribi (Figure 12.5).

6. *Rhino* (Figure 12.6).

7. *Kudu* (Figure 12.7).

8. *Bushpig*: Predominantly nocturnal and difficult to spot. Expensive techniques, such as radio telemetry are required to estimate abundance. In some cases, where bushpig are not hunted, line transects may be used.

9. *Sitatunga* (Figure 12.7).

10. *Hippos and Crocodiles* (Figure 12.8).

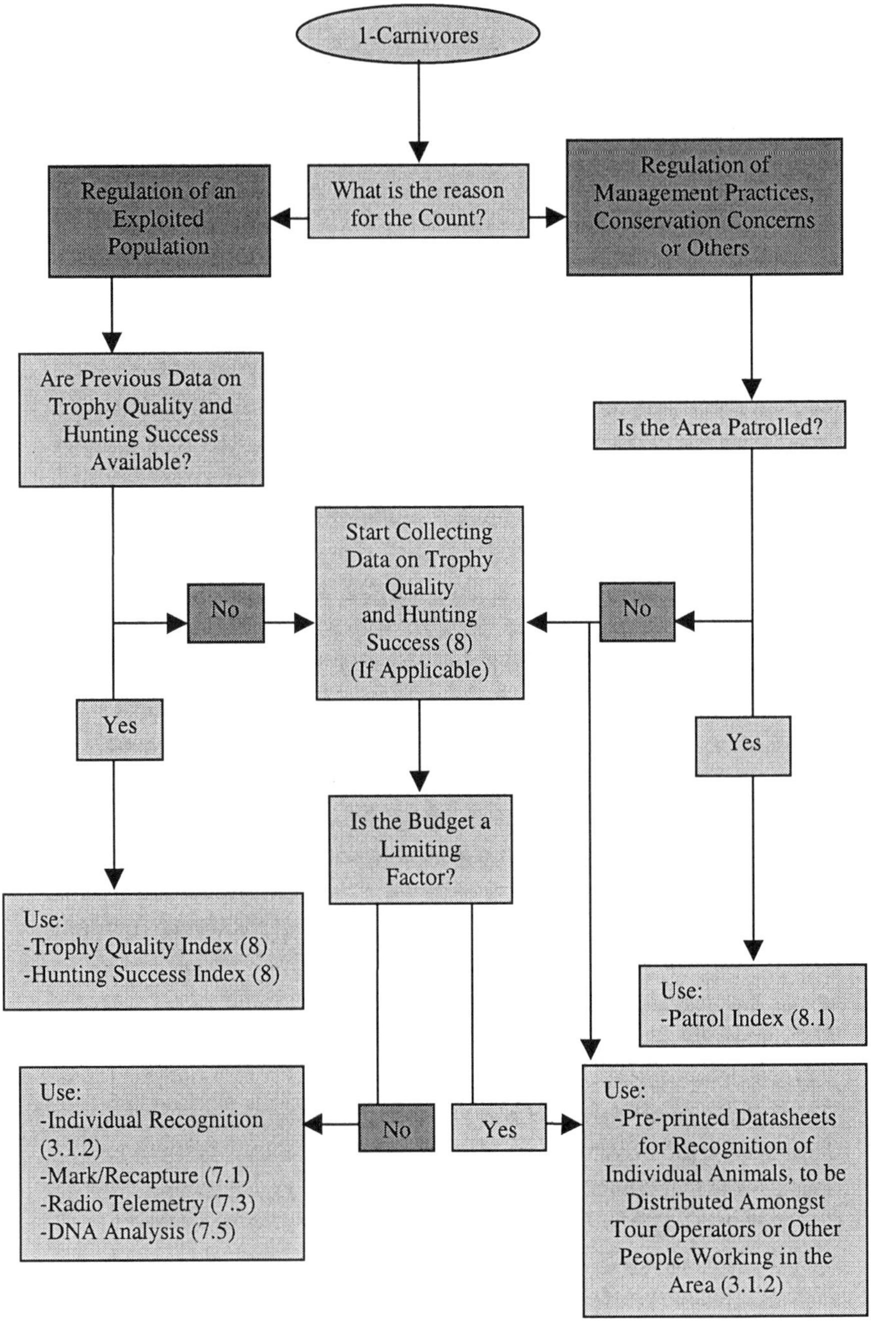

Figure 12.1: The sequence of decisions by which an indirect technique is chosen to assess trends in populations of carnivores. Numbers refer to chapters and sections describing the technique.

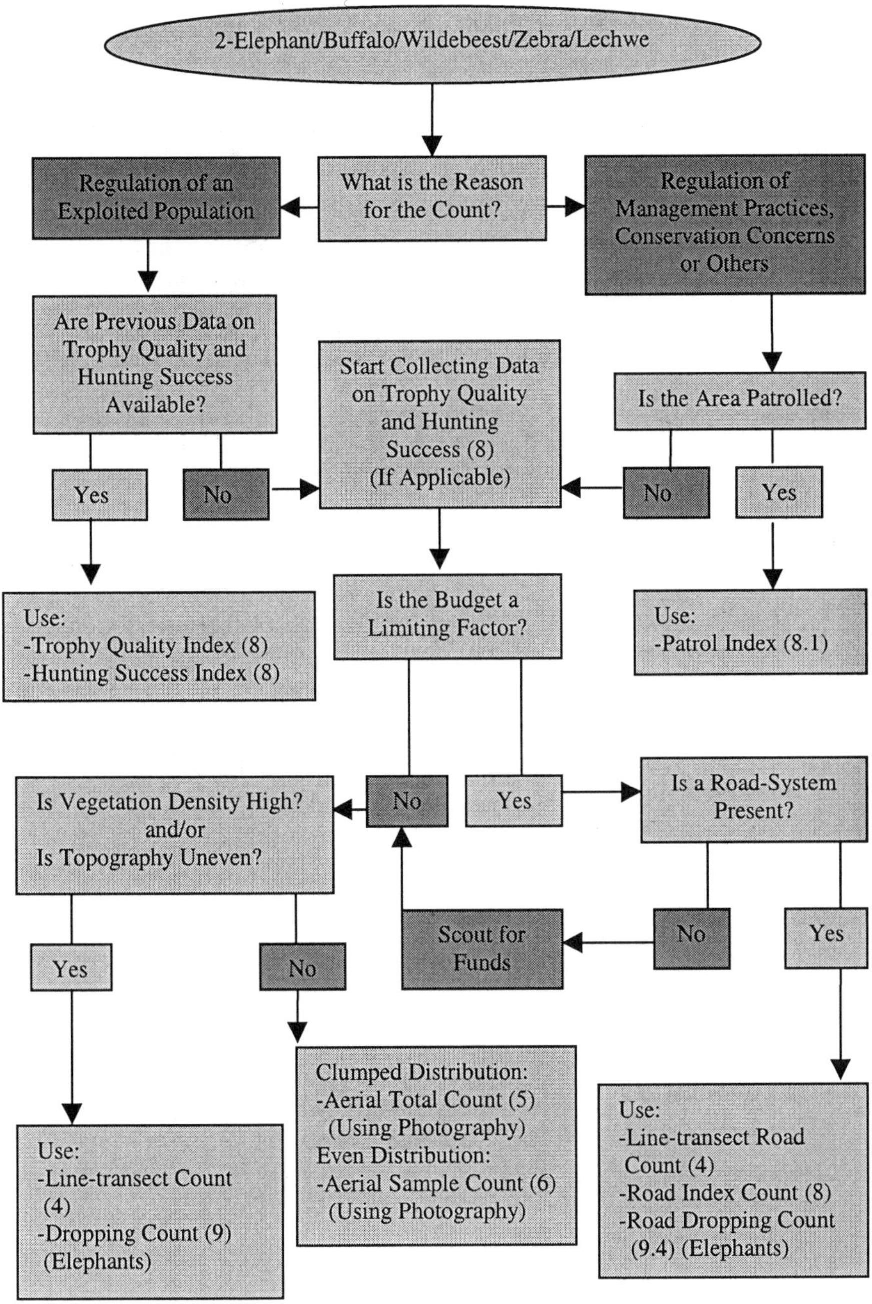

Figure 12.2: The sequence of decisions by which a technique is chosen to assess abundance of elephant, buffalo, wildebeest, zebra, and lechwe.

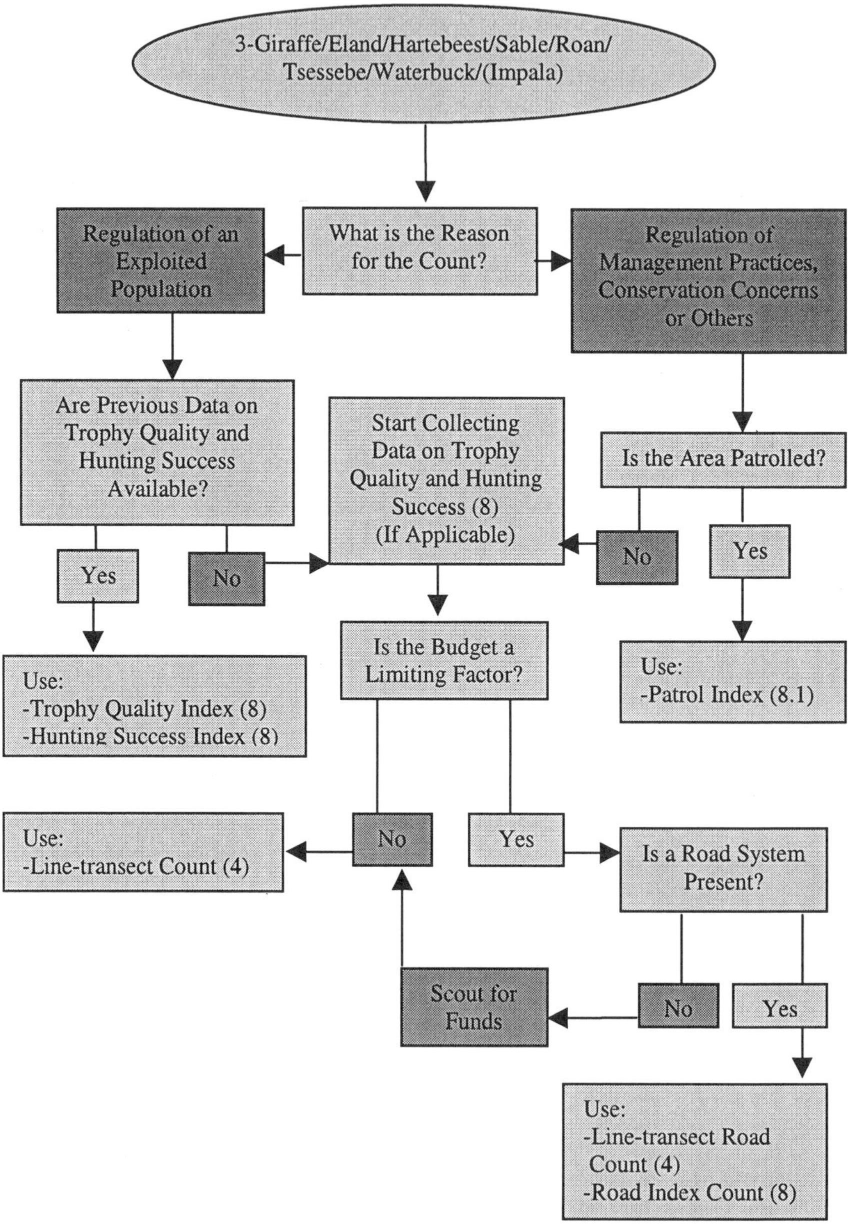

Figure 12.3: The sequence of decisions by which a technique is chosen to assess abundance of giraffe, eland, hartebeest, sable, roan, tsessebe, waterbuck and (impala).

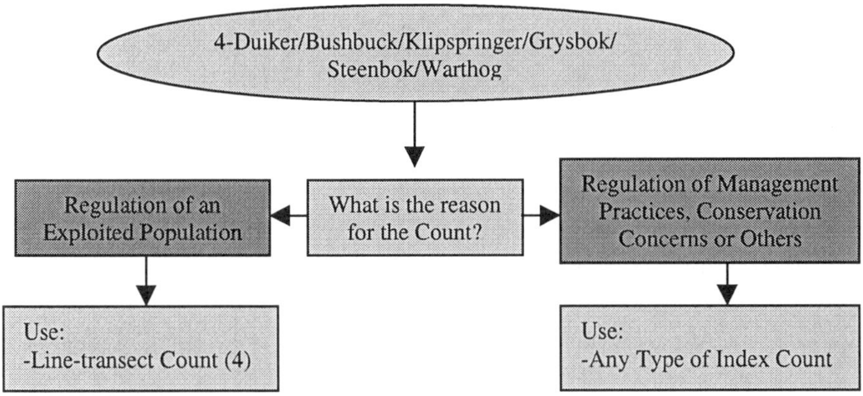

Figure 12.4: The sequence of decisions by which a technique is chosen to assess abundance of duiker, bushbuck, klipspringer, grysbok, steenbok and warthog.

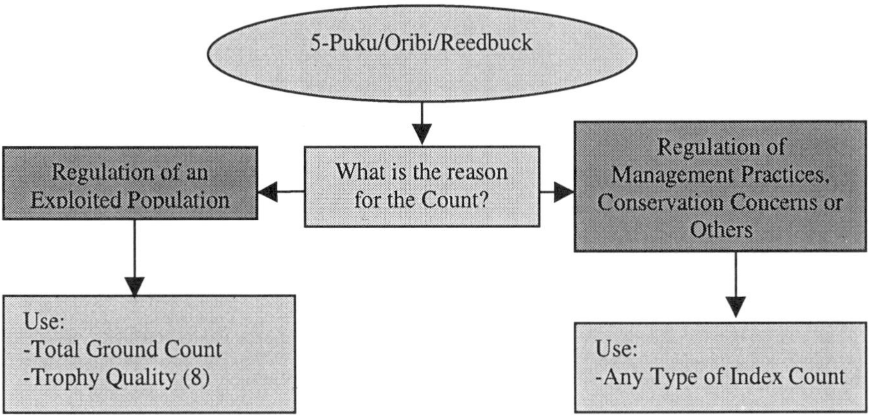

Figure 12.5: The sequence of decisions by which a technique is chosen to assess abundance of puku, oribi and reedbuck.

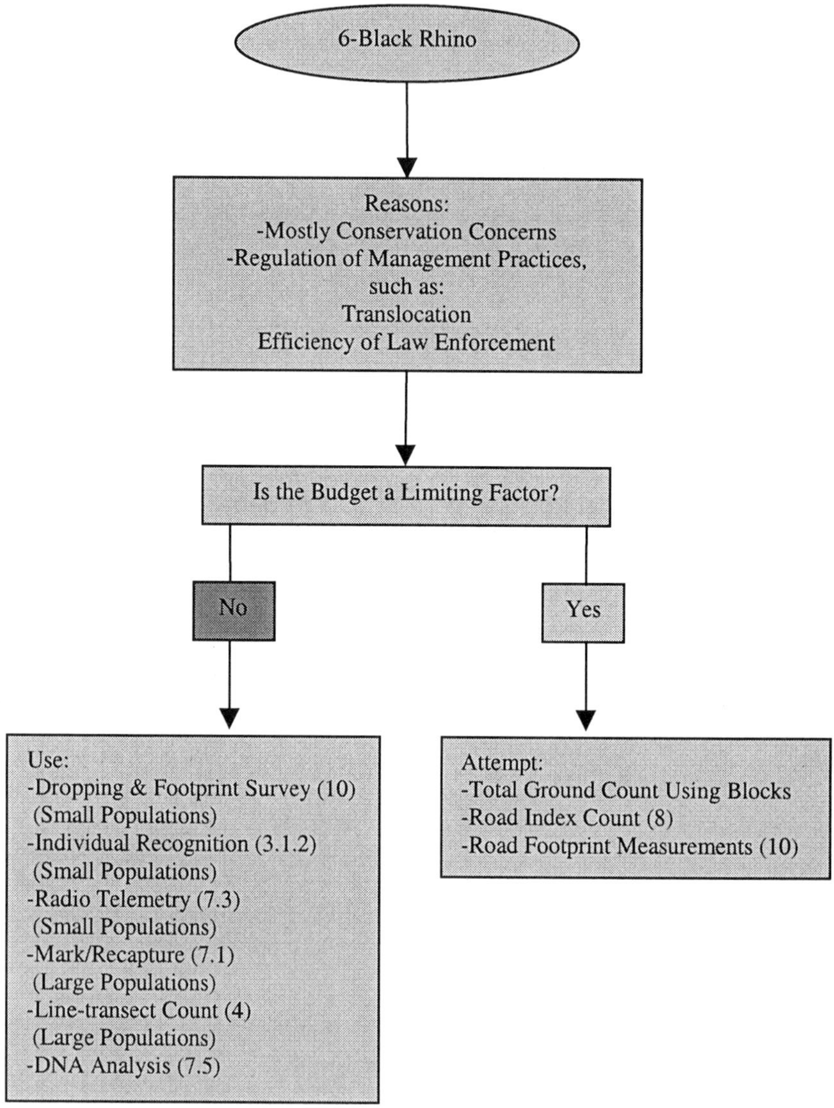

Figure 12.6: The sequence of decisions by which a technique is chosen to assess abundance of black rhino.

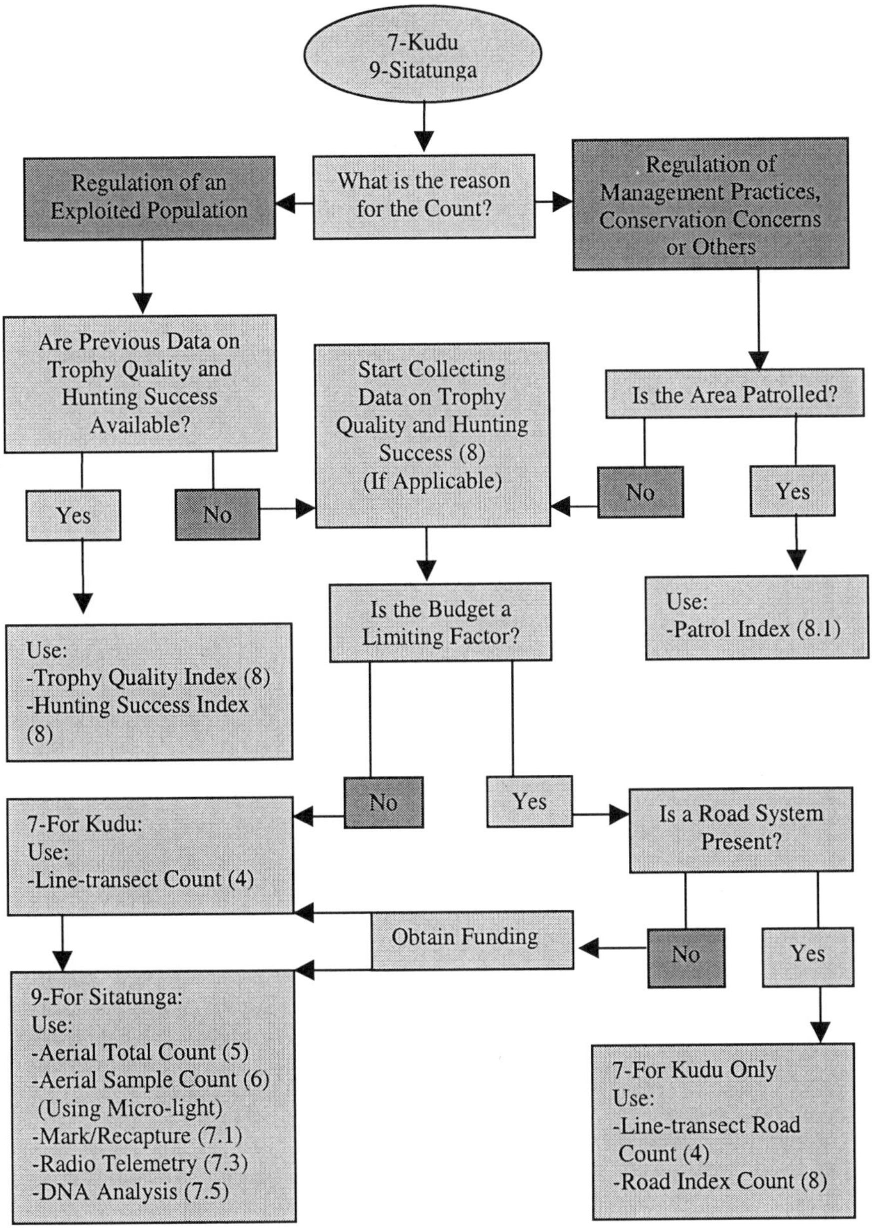

Figure 12.7: The sequence of decisions by which a technique is chosen to assess abundance of kudu and sitatunga.

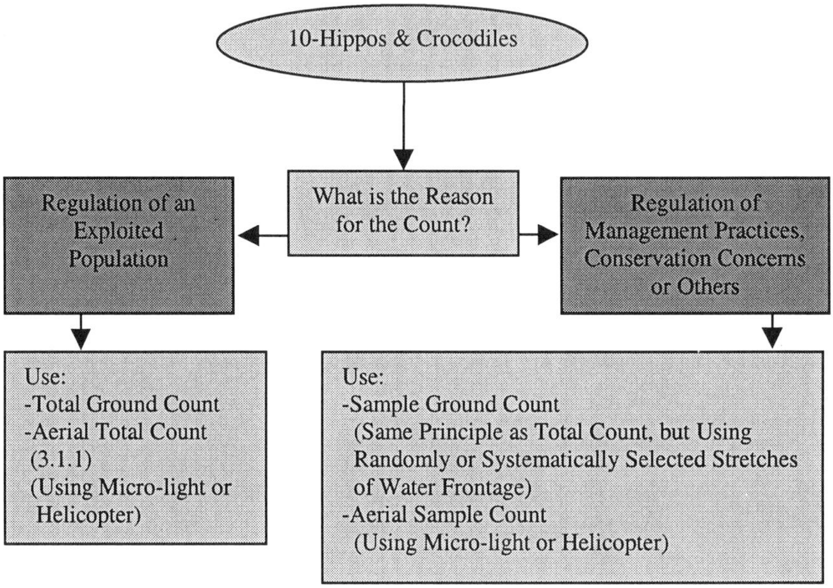

Figure 12.8: The sequence of decisions by which a technique is chosen to assess abundance of hippos and crocodiles.

References

Alers, M.P.T., Blom, A., Sikubwabo Kiyengo, C., Masunda, T. and R.F.W. Barnes (1992) Preliminary assessment of the status of the forest elephant in Zaïre. *Afr. J. Ecol.* **30,** 279 - 291.

Anderson, D.J. (1980) The home-range: a new nonparametric estimation technique. *Ecology* **63**, 103 – 112.

Anderson, D.R. and R.S. Pospahala (1970) Correction of bias in belt transects of immobile objects. *J. Wildl. Manage.* **34(1)**, 141 - 146.

Andrewartha, H.G. (1961) *Introduction to the Study of Animal Populations.* Chapman & Hall, London.

Bailey, N.T.J. (1981) *Statistical Methods in Biology.* Hodder and Stoughton, London.

Barnes, A., Hill, G.J.E. and G.R. Wilson (1986) Correcting for incomplete sighting in aerial surveys of kangaroos. *Aust. Wildl. Res.* **13**, 339 - 348.

Barnes, R.F.W. (1982) Elephant feeding behaviour in Ruaha National Park, Tanzania. *Afr. J. Ecol.* **20**, 123 – 136.

Barnes, R.F.W. (1993) Indirect methods for counting elephants in forest. *Pachyderm* **16**, 24 – 30.

Barnes, R.F.W. (1995) The great African dung rush. *Swara* **18(4)**, 24 - 25.

Barnes, R.F.W. (in press) How reliable are dung counts for estimating elephant numbers? *Afr. J. Ecol.*

Barnes, R.F.W. and K.L. Jensen (1987) How to count elephants in forest. *IUCN African Elephant and Rhino Specialist Group Technical Bulletin* **1**, 1 - 6.

Barnes, R.F.W., Barnes, K.L., Alers, M.P.T. and A. Blom (1991) Man determines the distribution of elephants in the rain forests of northeastern Gabon. *Afr. J. Ecol.* **29**, 54 - 63.

Barnes, R.F.W. and K.L. Barnes (1992) Estimating decay rates of elephant dung-piles in forest. *Afr. J. Ecol.* **30**, 316 - 321.

Barnes, R.F.W., Asamoah-Boateng, B., Naada Majam, J., Agyei-Ohemeng, J., Tchamba, M.N., Ekobo, A. and A. Nchanji (1994) Improving the accuracy of forest elephants census methods: studies of dung decay rates in Ghana and Cameroun. In: *African Elephant Conservation Programme Final Report*, Vol. **5**. The Environment & Development Group, Oxford.

Barnes, R.F.W., Blom, A., Alers, M.P.T. and K.L. Barnes (1995) An estimate of the numbers of forest elephants in Gabon. *J. Trop. Ecol.* **11**, 27 – 37.

Barnes, R.F.W., Beardsley, K., Michelmore, F., Barnes, K.L., Alers, M.P.T. and A. Blom (1997a) Estimating forest elephant numbers with dung counts and a geographic information system. *J. Wildl. Manage.* **61(4),** 1384 – 1393.

Barnes, R.F.W., Asamoah-Boateng, B., Naada Majam, J. and J. Agyei-Ohemeng (1997b) Rainfall and the population dynamics of elephant dung-piles in the forests of southern Ghana. *Afr. J. Ecol.* **35**, 39 - 52.

Barnes, R.F.W., Craig, G.C., Dublin, H.T., Overton, G., Simons, W. and C.R. Thouless (1999) *African Elephant Database 1998.* IUCN/SSC African Elephant Specialist Group. IUCN, Gland, Switzerland and Cambridge, UK.

Bell, R.H.V. (1981) An estimate of nyala numbers in Lengwe National Park by estimating the drinking rates of naturally recognisable animals. Dept. of National Parks and Wildlife, Lilongwe, Malawi.

Bell, R.H.V. and H.M. Banda (1983) Estimates of nyala numbers in Lengwe National Park. Dept. of National Parks and Wildlife, Lilongwe, Malawi.

Bell, R.H.V. and E. McShane-Caluzi (Eds.) **(1985)** *Conservation and Wildlife Management in Africa.* US Peace Corps, Washington, USA.

Boshe, J.T. and E.A. Lyimo (1983) Defecation rates, pellet and pellet group sizes of Kirk's dikdik in Arusha National Park, Tanzania. *Afr. J. Ecol.* **21**, 119 - 121.

Boulanger, J.G. and G.C. White (1990) A comparison of home-range estimators using Monte Carlo simulation. *J. Wildl. Manage.* **54(2)**, 310 – 315.

Buckland, S.T., Anderson, D.R., Burnham, K.P. and J.L. Laake (1993*)* *Distance Sampling: Estimating Abundance of Biological Populations.* Chapman & Hall, London.

Burnham, K.P., Anderson, D.R. and J.L. Laake (1980) Estimation of density from line-transect sampling of biological populations. *Wildl. Monogr.* **72**, 202pp.

Caughley, G.J. (1974) Bias in aerial surveys. *J. Wildl. Manage.* **38**, 921 - 933.

Caughley, G.J. (1977) *Analysis of Vertebrate Populations.* John Wiley & Sons, New York.

Caughley, G.J., Sinclair, R. and D. Scott-Kemis (1976) Experiments in aerial survey. *J. Wildl. Manage.* **40**, 290 - 300.

Caughley, G.J. and D. Grice (1982) A correction factor for counting emus from the air and its application to counts in Western Australia. *Aust. Wildl. Res.* **9**, 253 - 259.

Caughley, G.J. and A.R.E. Sinclair (1994) *Wildlife Ecology and Management.* Blackwell Science, Cambridge, USA.

Cochran, W.G. (1977) *Sampling Techniques.* John Wiley & Sons, New York.

Coe, M. (1972) Defecation by African elephants (*Loxodonta africana africana* (Blumenbach)). *E. Afr. Wildl. J.* **10**, 165 - 174.

Coe, M. (1977) The role of termites in the removal of elephant dung in Tsavo East National Park, Kenya. *E. Afr. Wildl. J.* **15**, 49 - 50.

Dawson, S. (1990) *A Model to Estimate Density of Asian Elephants (Elephas maximus) in Forest Habitats.* MSc. Thesis, University of Oxford.

Dawson, S. and A.J.F.M. Dekker (1992) Counting Asian elephants in forests. FAO, Bangkok.

Dixon, K.R. and J.A. Chapman (1980) Harmonic mean measure of animal activity areas. *Ecology* **61**, 1040 – 1044.

Dublin, H.T. (1986) *Decline of the Mara Woodlands: the Role of Fire and Elephants.* Ph.D. Thesis, University of British Columbia.

Emslie, R.H. (1993) *RHINO version 1.2/1.21. A population package designed for black rhinoceros in particular, but applicable to other species.* In collaboration with L.G. Underhill, H.J. van Hensbergen, K. Adcock, N. Pendock and W. Zucchini. Ecoscot Consultancy Services.

Fay, J.M. (1991) An elephant (*Loxodonta africana*) survey using dung counts in the forests of the Central African Republic. *J. Trop. Ecol.* **7**, 25 - 36.

Fay, J.M. and M. Agnagna (1991) A population survey of forest elephants (*Loxodonta africana cyclotis*) in northern Congo. *Afr. J. Ecol.* **29**, 177 - 187.

Foster, J.B. and D. Kearney (1967) Nairobi National Park game census, 1966. *E. Afr. Wildl. J.* **5**, 112 – 120.

Galli, N.S. and R.B. Flamand (1995) Darting and marking black rhinoceros on foot: part of a monitoring and population estimation technique in Hluhluwe-Umfolozi National Park, South Africa. *Pachyderm* **20**, 33 – 38.

Gibson, D.St.C. and G.C. Graig (1998) Trends of the elephant population in northern Botswana from aerial survey data. *Pachyderm* **25**, 14 – 27.

Goddard, J. (1967) The validity of censusing black rhinoceros populations from the air. *E. Afr. Wildl. J.* **5(1)**, 18 - 23.

Goodman, D. (1984) Statistics of reproductive rate estimates, and their implications for population projection. *Report of the International Whaling Commission Special Issue* **6**, 161 - 173.

Graham, A. and R.H.V. Bell (1989) Investigating observer bias in aerial survey by simultaneous double-counts. *J. Wildl. Manage.* **53(4)**, 1009 - 1016.

Grier, J.W., Gerrard, J.M., Hamilton, G.D. and P.A. Gray (1981) Aerial visibility bias and survey techniques for nesting bald eagles in northwestern Ontario. *J. Wildl. Manage.* **45**, 83 - 92.

Guy, P.R. (1975) The daily food intake of African elephant, *Loxodonta africana* Blumenbach, in Rhodesia. *Arnoldia* **7**, 1 – 6.

Harris, S., Cresswell, W.J., Forde, P.G., Trewhella, W.J., Woollard, T. and S. Wray Home range analysis using radio-tracking data – a review of problems and techniques particularly as applied to the study of mammals. *Mammal Rev.* **20,** 97 – 123.

Jachmann, H. (1984a) Status of the Mwabvi rhino (*Diceros bicornis*). *Nyala* **10(2)**, 77 - 90.

Jachmann, H. (1984b) Assessment of elephant numbers by means of dropping counts on tracks and its use in Nkhotakota Game Reserve, Malawi. *Nyala* **10(1)**, 33 - 38.

Jachmann, H. (1984c) *The Ecology of the Kasungu Elephant*. Ph.D. Thesis, University of Groningen.

Jachmann, H. (1988a) Comparison of road and ground surveys of large mammals at the Nazinga Game Ranch. ADEFA, Ouagadougou, Burkina Faso.

Jachmann, H. (1988b) Numbers, distribution and movements of the Nazinga elephants. *Pachyderm* **10**, 16 - 21.

Jachmann, H. (1989) Nombre, rêpartition et dêplacements des élèphants de Nazinga. *Nature et Faune* **5**, 25 - 35.

Jachmann, H. (1991a) Evaluation of four survey methods for estimating elephant densities. *Afr. J. Ecol.* **29**, 188 - 195.

Jachmann, H. (1991b) Current status of the Gourma elephants in Mali: a proposal for an integrated resource management project. IUCN, Gland, Switzerland.

Jachmann, H. (1995a) Survey experiments and aerial survey of elephants in the South Luangwa National Park and the Lupande Game Management Area, Zambia. *Pachyderm* **19**, 81 – 86.

Jachmann, H. (1995b) Manual to Program 'Aerial'. SLAMU, Chipata, Zambia.

Jachmann, H. (1996a) Line-transect sampling in the Lupande Game Management Area. SLAMU, Chipata, Zambia.

Jachmann, H. (1996b) Direct counts of elephants from the ground. In: *Studying Elephants* (ed. K. Kangwana). *African Wildlife Foundation Technical Handbook* No. **7**, AWF, Nairobi.

Jachmann, H. (1998) *Monitoring Illegal Wildlife Use and Law Enforcement in African Savanna Rangelands. WRMU Handbook # 1,* ECZ, Lusaka, Zambia.

Jachmann, H. (2000) *Zambia's Wildlife Resources: A Brief Ecology. WRMU Handbook # 2*, ECZ, Lusaka, Zambia.

Jachmann, H. and R.H.V. Bell (1979) The assessment of elephant numbers and occupance by means of dropping counts in the Kasungu National Park, Malawi. *Afr. J. Ecol.* **17**, 231 – 239.

Jachmann, H. and R.H.V. Bell (1984) The use of elephant droppings in assessing numbers, occupance and age-structure; a refinement of the method. *Afr. J. Ecol.* **22**, 127 – 141.

Jachmann, H. and T. Croes (1991) Effects of browsing by elephants on the *Combretum/ Terminalia* woodland at the Nazinga Game Ranch, Burkina Faso, West Africa. *Biol. Cons.* **57**, 13 –24.

Jachmann, H. and C.M. Phiri (1999a) Kasanka National Park; 1999 aerial sample counts, total counts, index counts and patrol report analysis. *WRMU Survey Report #* **3**. ECZ, Lusaka, Zambia.

Jachmann, H. and C.M. Phiri (1999b) Aerial survey of the central Luangwa Valley, animal abundance, distribution and trends. WRMU Survey Report # **2**. ECZ, Lusaka, Zambia.

Jennrich, R.I. and F.B. Turner (1969) Measurement of non-circular home range. *J. Theoretical Biol.* **22**, 227 – 237.

Jolly, G.M. (1969) Sampling methods for aerial censuses of wildlife populations. *E. Afr. Agric. For. J.* **34**, 46 – 49.

Jolly, G.M. and R.M. Watson (1979) Aerial sample survey methods in the quantitative assessment of ecological resources. In: *Sampling Biological Populations* (Eds. R.M. Gormack, G.P. Patib and D.S. Robson). Internal Co-operative Publishing House, Fairland, USA.

Kelly, H. and J. Beer (1996) Elephant status and ecology. Report on Kasanka National Park, Manchester Metropolitan University, Manchester, UK.

Koeppl, J.W., Slade, N.A. and R.S. Hoffman (1975) A bivariate home-range model with possible application to ethological data analysis. *J. Mammal.* **56**, 81 – 90.

Kohn, M.H., York, E.C., Kamradt, D.A., Haught, G., Sauvajot, R.M. and R.W. Wayne (1999) Estimating population size by genotyping faeces. *Proc. R. Soc. Lond.* **266,** 657-663

Kohn, M.H., and R.W. Wayne (2000) Facts from feces revisited. *Trends in Ecology and Evolution.* **12(6),** 223-227

Koster, S.H. and J.A. Hart (1988) Methods of estimating ungulate populations in tropical Forest. *Afr. J. Ecol.* **26**, 117 – 126.

Laake, J.L., Buckland, S.T., Anderson, D.R. and K.P. Burnham (1994) *Distance User's Guide.* Colorado Cooperative Fish & Wildlife Research Unit. Colorado State University, Fort Collins, CO.

Lamprey, H.F. (1963) Ecological separation of the large mammal species in the Tarangire Game Reserve, Tanganyika. *E. Afr. Wildl. J.* **1**, 1 – 30.

Laws, R.M., Parker, I.S.C. and R.C.B. Johnstone (1975) *Elephants and their Habitats.* Clarendon Press, Oxford, UK.

Lebreton, J.D., Burnham, K.P., Clobert, J. and D.R. Anderson (1992) Modelling survival and testing biological hypotheses using marked animals: a unified approach with case studies. *Ecol. Mon.* **62**, 67 – 118.

Lewis, D.M. (1986) Disturbance effects on elephant feeding: evidence for compression in Luangwa Valley, Zambia. *Afr. J. Ecol.* **24**, 227 – 241.

Macdonald, D.W. and C.J. Amlaner (1980) A Practical Guide to Radio Tracking. In: *A Handbook on Biotelemetry and Radio Tracking* (Eds. C.J. Amlaner and D.W. Macdonald). Pergamon Press, Oxford, UK.

Magin, C.D. (1989) Variability in total ground counts on a Kenyan Game Ranch. *Afr. J. Ecol.* **27**, 297 – 303.

Magnusson, W.E., Caughley, G.J. and G.C. Grigg (1977) A double survey estimate of population size from incomplete counts. *J.Wildl. Manage.* **42(1)**, 174 – 176.

Martin, R.B., Craig, G.C. and V.R. Booth (Eds) **(1992)** *Elephant Management in Zimbabwe.* Dept of National Parks and Wildlife Management, Harare, Zimbabwe.

Mason, D.R. (1990) Monitoring of sex and age ratios in ungulate populations of the Kruger National Park by ground survey. *Koedoe* **33**, 19 – 28.

McClanahan, T.R. (1986) Quick population survey method using faecal droppings and a steady state assumption. *Afr. J. Ecol.* **24**, 37 – 39.

Merz, G. (1986) Counting elephants (*Loxodonta africana cyclotis*) in tropical rainforests with particular reference to the Tai National Park, Ivory Coast. *Afr. J. Ecol.* **24**, 61 – 68.

Moss, C.J. and J.H. Poole (1983) Relationships and social structure in African elephants. In: *Primate Social Relationships: An Integrated Approach.* (Ed. R.H. Hinde). Blackwell Scientific Publications, Oxford, UK.

Mowat, G. and C. Strobeck (2000) Estimating population size of grizzly bears using hair capture, DNA profiling and mark/recapture analysis. *J. Wildl. Manage* **64(1),** 183-193.

Mphande, J.N.B. and H.S. Jamusana (1984) Culling of nyala antelopes in Lengwe National Park, Malawi. In: *Conservation and Wildlife Management in Africa* (Eds. R.H.V. Bell and E. McShane-Caluzi). US Peace Corps, Washington, USA.

Neff, D.J. (1968) The pellet-group count technique for big game trend, census and distribution: a review. *J. Wildl. Manage.* **32**: 597 – 614.

Norton-Griffiths, M. (1975) *Counting Animals* (1st Edition, Ed. J.J.R. Grimsdell). *African Wildlife Foundation Technical Handbook* No. **1**. AWF, Nairobi, Kenya.

Norton-Griffiths, M. (1978) *Counting Animals* (2nd Edition, Ed. J.J.R. Grimsdell). *African Wildlife Foundation Technical Handbook* No. **1**. AWF, Nairobi, Kenya.

Norton-Griffiths, M. and H.T. Santibanez (1980) Evaluation of ground and aerial census work on vicuna in Pampa Galeras, Peru. IUCN and WWF Publication, Gland, Switzerland.

Pace, R.M. and H.P. Weeks (1990) A non-linear weighted least-squares estimator for radio-tracking via triangulation. *J. Wildl. Manage.* **54(2)**, 304 – 310.

Parker, I.S.C. (1976) Black rhinoceros and other large mammals in the Mwabvi Game Reserve. Dept. of National Parks and Wildlife, Lilongwe, Malawi.

Plumptre, A.J. and S. Harris (1995) Estimating the biomass of large mammalian herbivores in a tropical montane forest: a method of faecal counting that avoids assuming a steady state system. *J. Appl. Ecol.* **32**, 111 –120.

Ridding, C. (1975) Mwabvi Game Reserve. Dept. of National Parks and Wildlife, Lilongwe, Malawi.

Said, M.Y., Chunge, R.N., Craig, G.C., Thouless, C.R., Barnes, R.F.W. and H.T. Dublin (1995) *African Elephant Database 1995*. IUCN, Gland, Switzerland.

Seber, G.A.F. (1982) *The Estimation of Animal Abundance*. Charles Griffin, London, UK.

Seber, G.A.F. (1986) A review of estimating animal abundance. *Biometrics* **42**, 267 – 292.

Seber, G.A.F. (1992) A review of estimating animal abundance. *Int. Stat. Review* **60**, 129 – 166.

Short, J. (1983) Density and seasonal movements of forest elephant (*Loxodonta africana cyclotis*, Matschie) in Bia National Park, Ghana. *Afr. J. Ecol.* **21**, 175 – 184.

Smallwood, K.S. and E.L. Fitzhugh (1993) A rigorous technique for identifying individual mountain lions *Felis concolor* by their tracks. *Biol. Cons.* **65**, 51 – 59.

Sukumar, R., Varman, K.S., Santosh, J.A. and U. Ramakrishnan (1991) Methods of estimating elephant numbers: a review. In: *Censusing elephants in forests: Proceedings of an International Workshop* (Eds. U. Ramakrishnan, J.A. Santosh and R. Sukumar). Asian Elephant Conservation Centre, Bangalore, India.

Tchamba, M. (1992) Defecation by the African forest elephant (*Loxodonta africana cyclotis)* in the Santchou Reserve, Cameroun. *Mammalia* **56,** 155-158.

Van Dyke, F.G., Brocke, R.H. and H.G. Shaw (1986) Use of road track counts as indices of mountain lion presence. *J. Wildl. Manage.* **50**, 102 – 109.

Voigt, D.R. and R.R. Tinline (1980) Strategies for analysing radio tracking movement data. In: *A handbook on Biotelemetry and Radio Tracking* (Eds. C.J. Amlaner and D.W. Macdonald). Pergamon Press, Oxford, UK.

Watson, R.M., Jolly, G.M. and A.D. Graham (1968) Two experimental censuses. *E. Afr. Agr. For. J.* **34**, 60 – 62.

Watson, R.M. and R.H.V. Bell (1969) The distribution, abundance and status of elephant in the Serengeti region of northern Tanzania. *J. Appl. Ecol.* **6**, 115 – 132.

Weir, J.S. (1971) The effects of the creation of additional water supplies on a Central African National Park. In: *Scientific Management of Animal and Plant Communities for Conservation* (Eds. E. Duffey and A.S. Watt). Blackwell Scientific Publications, Oxford, UK.

Western, D. and J.J.R. Grimsdell (1979) *Measuring the Distribution of Animals in Relation to the Environment.* AWF Technical Handbook No. **2**, AWF, Nairobi, Kenya.

Western, D., Moss, C.J. and N. Georgiadis (1983) Age estimation and population age structure of elephants from footprint dimensions. *J. Wildl. Manage.* **47(4)**, 1192 – 1197.

White, L.J.T. (1995) Factors affecting the duration of elephant dung piles in rain forest in the Lope Reserve, Gabon. *Afr. J. Ecol.* **33**, 142 – 150.

Whyte, I.J. (1996) Studying elephant movements. In: *Studying Elephants* (Ed. K. Kangwana). AWF Technical Handbook No. **7**, AWF, Nairobi, Kenya.

Wing, L.D. and I.O. Buss (1970) Elephants and forests. *Wildl. Monogr.* **19**, 1 – 92.

Wyatt, J.R. and S.K. Eltringham (1974) The daily activity of the elephant in the Rwenzori National Park, Uganda. *E. Afr. Wildl. J.* **12**(**4**), 273 – 290.

Yamazaki, K. (1992) Social variation of lions in a male depopulated area in the Luangwa Valley, Zambia. SLAMU, Chipata, Zambia.

APPENDICES

Appendix I

A BRIEF MANUAL TO PROGRAM "AERIAL"

Introduction

"AERIAL" is a user-friendly computer program, produced to analyse aerial survey data to estimate animal abundance. The program is created mainly to analyse survey data using transects, but it can also be used for blocks of the same size. AERIAL runs on all computers with DOS and an EGA or (S)VGA Video Adapter.

To run the program, start the computer in DOS mode and type "AERIAL" at the prompt. The program begins with a brief explanation, after which the "MAIN MENU" appears with 7 different options.

Before you start a particular analysis, make sure that you have compiled a list of the information required by the program. For the particular data required, consult the relevant paragraph below. The program can only deal with one animal species at the time, using a minimum of 2 and a maximum of 200 transects. You do not have to restart the program if you wish to analyse the data for more than one species, but make sure you copy the "RESULTS" page or use the "PRINT SCREEN" on your keyboard before continuing with the next run. The program will automatically erase previous data when you proceed with the next data set. When entering data, make sure that you check your input before pressing "ENTER". On a mistake, use "BSPACE" for correction, however, after the number has been entered, you will not have a second chance to correct a mistake, unless you start again! Certain data, such as the length of transects, may require decimal points to increase precision and accuracy. The program allows you to use decimal points. The program allows analysis of aerial sample data using the following methods:

Jolly's Method 1 For Equal Sized Sample Units

Jolly's Method 1 should be used when the transects are more or less of the same size. You do not need to know transect length nor the size of the survey area. Using this method, you will avoid bias introduced by inaccuracies in measuring transect lengths. However, when the transects are of different length, using this method will inflate sample variance.

With a GPS, the beginning and end of a transect can be determined accurately.

Before you enter the number of animals counted in each transect, the program will ask for the "NUMBER OF TRANSECTS" in the sample, the "BASE-LINE LENGTH" in km and the "TRANSECT WIDTH" in meters. Do not forget to sum left and right strips. When the survey included a double count, observer-bias correction factors can be entered. The "RESULTS" page will show the % Coverage, the Population Estimate, the Population Variance, the Standard Error, the 95% Confidence Limits and the % Confidence Limits.

The percentage results and the population variance are presented with decimal points, whereas the other results have been rounded. Also, the program automatically corrects the 95% confidence limits for sample size or number of transects sampled!

Jolly's Method 2 For Unequal Sized Sample Units

Using this method, the size of the survey area must be known, as well as the length of each transect. The program begins by asking you for the "SIZE OF THE SURVEY AREA" in km², the "NUMBER OF TRANSECTS" in the sample and the "TRANSECT OR STRIP-WIDTH" in meters. Before you enter the data, prepare a list of all transects, the length (km) and numbers of animals of each species counted. The program first asks for the length of each transect and will then continue with the number of animals counted in each transect. The "OUTPUT" is the same as that for Jolly's Method 1.

Combining Results From a Stratified Sample Count

With a stratified count, each stratum is treated as a separate area. The program can be used to combine the results for each stratum to give a population estimate for the entire survey area.

On the first page, the program will ask you for the "NUMBER OF STRATA" the "POPULATION ESTIMATES FOR EACH STRATUM", the "POPULATION VARIANCE FOR EACH STRATUM" and the "NUMBER OF TRANSECTS FLOWN IN EACH STRATUM". Make sure you number the strata and enter the data in the same order. The "RESULTS" page is the same as that for the two methods described previously.

Comparing Two Population Estimates

This option can be used to investigate whether the size of a population changed over time, or whether the results of two different methods are identical. The test provided in this program can only be used if the sample size or the number of transects flown is more than 30!

The program will ask you for the "POPULATION ESTIMATES" and the "POPULATION VARIANCES". The result will either be insignificant (the estimates are the same) or significant at the 5% level ($P< 0.05$).

Merging Two Or More Independent Population Estimates

If biases are kept constant, two or more estimates of the same population can be used to arrive at a single estimate with low variance. However, one must first investigate whether any differences between population estimates are due to sampling error alone or to real differences, such as migration. *A VARIANCE RATIO or F-TEST* is applied to the data, and only when the result is insignificant, the estimates will be merged. A maximum of 10 population estimates can be merged.

On the first page, the program will ask for the "NUMBER OF ESTIMATES" to be merged, the "POPULATION ESTIMATES" and the "POPULATION VARIANCES". On the bottom of the same page, the program will indicate whether the F-Ratio is significant (population estimates will not be merged!) or insignificant (population estimates will be merged). The results of the merged population estimates are presented in the same way as with previous methods.

Appendix II

PROGRAM "AERIAL" IN QUICKBASIC 4.5

```
CLS
CLEAR , , 5000
KEY 1, "AO": KEY 2 , "B1": KEY 3, "C2": KEY 4, "D3": KEY 5 , "E4"
KEY 6, "F 5": KEY 7, "G6": KEY 8, "H7": KEY 9, "I8": KEY 10, "J 9"
15  :    SCREEN 1
CLS
COLOR 1, 6
XCent = 160: YCent = 100
LINE (XCent - 50 , YCent - 50) - (XCent + 50 , YCent + 50) , 3 , B
LINE (XCent - 60 , YCent - 60) - (XCent + 60 , YCent + 60) , 3 , B
LINE (XCent - 70 , YCent - 70) - (XCent + 70 , YCent + 70) , 3 , B
LINE (XCent - 80 , YCent - 80) - (XCent + 80 , YCent + 80) , 3 , B
LOCATE 10, 18
PRINT "AERIAL"
LOCATE 12 , 18
PRINT "SURVEY"
LOCATE 14, 17
PRINT "ANALYSIS"
LOCATE 18, 16
PRINT "H. Jachmann"
LOCATE 24, 9
PRINT "Press any key to continue"
GOSUB 50
CLS
REM LOADING VARIABLES AND ARRAYS
REDIM TL (1 TO 200) AS SINGLE
REDIM Z(1 TO 200) AS SINGLE
REDIM y (1 TO 200) AS LONG
REDIM Z2 (1 TO 200) AS SINGLE
REDIM Zy (1 TO 200) AS SINGLE
REDIM V (1 TO 20) AS INTEGER
REDIM Q (1 TO 20) AS DOUBLE
REDIM P ( 1 TO 20) AS DOUBLE
REDIM POP (1 TO 10) AS SINGLE
REDIM VARI (1 TO 10) AS SINGLE
REDIM VG2 (1 TO 10) AS DOUBLE
```

```
REDIM W (1 TO 10) AS DOUBLE
REDIM M (1 TO 10) AS SINGLE
DIM SP1 AS STRING
DIM SP2 AS STRING
DIM AREA AS STRING
TUSSEN# = 0
Ty2# = 0
A = 0
B = 0
S = 0
CF = 0
number = 0
Ty = 0
PE = 0
VA# = 0
PCL = 0
VG1 = 0
VG2T = 0
K = 0
K1 = 0
K2 = 0
K3 = 0
K4 = 0
K5 = 0
K20 = 0
CO = 0
SE = 0
CL = 0
PE1 = 0
PE2 = 0
t = 0
t1 = 0
T2 = 0
t3 = 0
t4 = 0
t5 = 0
t6 = 0
t7 = 0
Tz = 0
Tz2# = 0
Tzy = 0
t20 = 0
```

```
Tp = 0
Tq = 0
Tv = 0
D = 0
P1 = 0
P2 = 0
P3 = 0
V1 = 0
V2 = 0
V3 = 0
VG = 0
PG = 0
SV = 0
W = 0
F = 0
MT = 0
CLS
SCREEN 8
LINE (50, 10) – (590, 190) , 3 , B
COLOR 4, 7
LOCATE 2, 27
PRINT "AERIAL SURVEY ANALYSIS"
LOCATE 5, 9
PRINT "This program analyses aerial survey data, using"
LOCATE 6 , 9
PRINT "Jolly's Method 1 or 2, for equal or unequal – sized"
LOCATE 7 , 9
PRINT "sample units. Depending on the method,"
LOCATE 8, 9
PRINT "the program will ask for the Size of the Survey"
LOCATE 9, 9
PRINT "Area in km², the number of transects, the Base-"
LOCATE 10 , 9
PRINT "Line in km (this is the maximum length of the survey"
LOCATE 11, 9
PRINT "area at right angles with the transects) and Transect Width"
LOCATE 12 , 9
PRINT "in m (add up left and right strips in meters)."
LOCATE 14, 9
PRINT "Before you start the program, prepare a list with the"
LOCATE 15 , 9
PRINT "information described above, the length of each transect"
```

```
LOCATE 16, 9
PRINT "and the numbers of each species observed in that particular"
LOCATE 17 , 9
PRINT "transect. The program can only deal with one species,"
LOCATE 18 , 9
PRINT "using a maximum of 200 transects. Before you"
LOCATE 19, 9
PRINT "start, number the transects from 1 to n. Also"
LOCATE 20, 9
PRINT "do not forget to include transects with 0 animals observed!"
LOCATE 24, 9
PRINT "Press any key to continue"
GOSUB 50
CLS
LINE (50, 10) – (590 , 190) , 3, B
COLOR 4 , 7
LOCATE 5 , 9
PRINT "With Jolly's Methods 1 and 2, and with Combining Strata you have"
LOCATE 6 , 9
PRINT "the option to enter the name of the conservation area and the"
LOCATE 7, 9
PRINT "name of the animal species for easy reference on the results"
LOCATE 8, 9
PRINT "page (when using the PRINT SCREEN key)!"
LOCATE 10, 9
PRINT "When the double-count procedure was followed,"
LOCATE 11 , 9
PRINT "observer bias correction factors will be available for each species."
LOCATE 12, 9
PRINT "The program allows you to enter these"
LOCATE 13 , 9
PRINT "correction factors. Press Y when a correction"
LOCATE 14, 9
PRINT "factor is available and N when it is not available)."
LOCATE 15 , 9
PRINT "The program will now revise all population statistics accordingly!"
LOCATE 24, 9
PRINT "Press any key to continue"
GOSUB 50
```

```
CLS
LINE (50, 10) – (590, 190) , 3 , B
COLOR 4, 7
LOCATE 5 , 9
PRINT "With a stratified design, each stratum should be dealt with"
LOCATE 6 , 9
PRINT "separately. The strata can be combined afterwards"
LOCATE 7 , 9
PRINT "to give a population estimate for the entire survey"
LOCATE 8 , 9
PRINT "area. Also, the program can compare and merge two or more"
LOCATE 9 , 9
PRINT "estimates."
LOCATE 11, 9
PRINT "Make sure you check your input before you continue"
LOCATE 12 , 9
PRINT "with the next figure. On a mistake, use backspace for correction."
LOCATE 13, 9
PRINT "You will not have a second chance to make a correction!"
LOCATE 14 , 9
PRINT "After you have typed a figure and checked it, use ENTER before"
LOCATE 15, 9
PRINT "you continue with the next figure."
LOCATE 24, 9
PRINT "Press any key to continue"
GOSUB 50
CLS
MENU:
ON ERROR GOTO erro
CLS
LINE (50, 10) – (590, 191) , 3 , B
COLOR 5, 0
LOCATE 2 , 33
PRINT "MAIN MENU"
LOCATE 4 , 9
PRINT "F1 = JOLLY'S METHOD 1 FOR EQUAL SIZED SAMPE
UNITS"
LOCATE 6 , 9
PRINT "F2 = JOLLY'S METHOD 2 FOR UNEQUAL SIZED SAMPLE
UNITS"
LOCATE 8 , 9
```

```
PRINT "F3 = COMBINING RESULTS FROM A STRATIFIED
SAMPLE COUNT"
LOCATE 10 , 9
PRINT "F4 = COMPARING TWO POPULATION ESTIMATES"
LOCATE 12, 9
PRINT "F5 = MERGING TWO OR MORE POPULATION
ESTIMATES"
LOCATE 14, 9
PRINT "F6 = HELP"
LOCATE 16, 9
PRINT "F7 = EXIT PROGRAM"
20 :    GOSUB 50
IF L4 = 1 THEN GOTO METHOD 1
IF L4 = 2 THEN GOTO METHOD 2
IF L4 = 3 THEN GOTO METHOD 3
IF L4 = 4 THEN GOTO METHOD 4
IF L4 = 5 THEN GOTO METHOD 5
IF L4 = 6 THEN GOTO HELP
IF L4 = 7 THEN GOTO FINISH
IF L4 > 7 THEN BEEP
IF L2 OR L3 THEN BEEP
GOTO 20

50 :    n$ = INKEY$: IF NOT n$ = " " THEN 50
51 :    n$ = INKEY$: IF n$ = " " THEN 51
        n1$ = INKEY$
        L = VAL (n$) : L1 = ASC (n$)
        IF L1 > 96 AND L1 < 123 THEN L2 = L1 - 32 ELSE L2 = L1
        L3 = 0: IF LEN (n$) > 1 THEN L$ = RIGHT$ (n$, 1) : L3 = ASC
        (L$)
        L4 = 0: IF LEN (n1$) = 0 THEN RETURN ELSE L4 = ASC
        (n1$) - 47
        RETURN
62 :    X = 0
        L$ = " "
        t20 = t20 + 1
63 :    LOCATE t20, 37: PRINT L$; " "
        GOSUB 50
        IF L2 = 8 AND LEN (L$) > 0 THEN L$ = LEFT$ (L$, LEN (L$)
        - 1)
        IF L2 = 13 THEN X = VAL (L$) : RETURN
        IF L2 = 46 THEN L$ = L$ + " . "
```

```
        IF L > 0 OR n$ = "0" THEN L$ = L$ + n$
        GOTO 63

64  :   SP1 = " "
        L$ = " "
        t20 = t20 + 1
65  :   LOCATE t20 , 37: PRINT L$; " "
        GOSUB 50
        IF L2 = 8 AND LEN (L$) > 0 THEN L$ = LEFT$ (L$ , LEN
        (L$) - 1)
        IF L2 > 64 AND L2 < 91 THEN L$ = L$ + CHR$ (L2)
        IF L2 = 13 THEN SP1 = L$: RETURN
        IF L > 0 OR n$ = "0" THEN L$ = L$ + n$
        GOTO 65

METHOD1:
                CLS
                B = 0
                S = 0
                CF = 0
                AREA = " "
                SP2 = " "
                number = 0
                t20 = 9
                SCREEN 8
                COLOR 7, 6
                LOCATE 4, 9
                PRINT "JOLLY'S METHOD 1 FOR EQUAL SIZED
                SAMPLE UNITS"
                LOCATE 5 , 9
                PRINT "------------------------------------------------------------"
                LOCATE 7 , 19
                PRINT "Use BSPACE for correction ! ! ! !"

101 :   LOCATE 10 , 9
        PRINT "Number of Transects = ? "
        GOSUB 62
        number = X
        IF X < 2 OR X > 200 THEN
        BEEP
        LOCATE 10 , 37
        PRINT "    "
```

```
        LOCATE 10, 45
        PRINT " > 1 AND < 200 transects"
        t20 = t20 - 1
        GOTO 101
        END IF
102:    LOCATE 11, 9
        PRINT "Base – line length (km) = ? "
        GOSUB 62
        B = X
        IF X = 0 THEN
        BEEP
        LOCATE 11, 45
        PRINT "     "
        t20 = t20 - 1
        GOTO 102
        END IF
103:    LOCATE 12, 9
        PRINT "Transect-width (m) = ?"
        GOSUB 62
        S = X / 1000
        IF X = 0 THEN
        BEEP
        LOCATE 12, 45
        PRINT "     "
        t20 = t20 - 1
        GOTO 103
        END IF

        LOCATE 13, 9
        PRINT "Conservation Area = ?"
        GOSUB 64
        AREA = SP1

        LOCATE 14, 9
        PRINT "Animal Species = ? "
        GOSUB 64
        SP2 = SP1

TUS1:   LOCATE 16, 9
        PRINT "DO YOU HAVE A CORRECTION FACTOR FOR
        THIS SPECIES,"
        LOCATE 17, 9
```

```
        PRINT "OBTAINED THROUGH A DOUBLE-COUNT
        PROCEDURE, (Y)ES OR (N) O?"
        GOSUB 50
        IF L2 = 78 THEN CF = 1: GOTO TUS5
        IF L2 = 89 THEN GOTO TUS3
        IF NOT L2 = 78 OR L2 = 89 THEN BEEP: GOTO TUS1

TUS3:   t20 = 18
        LOCATE 19, 9
        PRINT "Correction factor = ?"
        GOSUB 62
        CF = X
        IF X < 1 THEN
        BEEP
        LOCATE 19, 35
        PRINT "     "
        t20 = t20 - 1
        GOTO TUS3
        END IF

TUS5:   LOCATE 24, 9
        PRINT "Press any key to continue"
        GOSUB 50
104:    CLS
        t = 0
        t5 = 0
        t6 = 0
        t7 = 0
        Ty = 0
        Ty2# = 0
        K = 0
        K1 = 0
        K2 = 0
        K3 = 0
        K4 = 0
        K5 = 0
        K20 = 0
        CO = 0
        PCL = 0
        TUSSEN# = 0
        PE = 0
        PE1 = 0
```

```
        VA# = 0
        SE = 0
        CL = 0
        FOR t 8 = 1 TO number
        y(t8) = 0
        NEXT t 8
        t20 = 9
        LOCATE 2, 9
        PRINT "ENTER NUMBER OF ANIMALS OBSERVED IN
        EACH TRANSECT"
        LOCATE 4, 9
        PRINT "USE BSPACE FOR CORRECTION ! ! !""
        FOR t5 = 1 TO number
        LOCATE 10, 9
        PRINT "Transect number"; t5
        GOSUB 62
        y (t5) = X
        LOCATE 10, 37
        PRINT "     "
        t20 = t20 - 1
        NEXT t5
        LOCATE 24, 9
        PRINT "----------------------------------------------------------------"
        LOCATE 25, 9
        PRINT "Press any key to continue"
        GOSUB 50
105:    t = 0
        IF number < 6 THEN t = 2.6
        IF number < 8 AND number > 5 THEN t = 2.4
        IF number < 10 AND number > 7 THEN t = 2.3
        IF number < 13 AND number > 9 THEN t = 2.2
        IF number < 27 AND number > 12 THEN t = 2.1
        IF number > 27 THEN t = 1.96
109:    FOR t6 = 1 TO number
        Ty = Ty + y (t6)
        NEXT t6
        FOR t7 = 1 TO number
        TUSSEN# = y (t 7) © 2
        Ty2# = Ty2# + TUSSEN#
        TUSSEN# = 0
        NEXT   t 7
        K = B / S
```

```
        K1 = (Ty © 2) / number
        K2 = (K © 2 - K * number) / number
        K3 = 1 / (number - 1)
        K4 = Ty2# - K1
        K5 = K3 * K4
        K20 = (number * S) / B
        PE1 = K * (Ty / number)
        PE = CF * PE1
        VA# = (CF © 2) * (K2 * K5)
        SE = SQR (VA#)
        CL = t * SE
        PE = CLNG (PE)
        SE = CLNG (SE)
        CL = CLNG (CL)
        CO = K20 * 100
        PCL = (CL / PE) * 100
110:    CLS
        LOCATE 4, 35
        PRINT "RESULTS"
        LOCATE 5, 35
        PRINT "-------"
        LOCATE 8, 9
        PRINT "CONSERVATION AREA = " : AREA
        LOCATE 10, 9
        PRINT "ANIMAL SPECIES = "; SP2
        LOCATE 12 , 9
        PRINT "CORRECTION FACTOR ="; CF
        LOCATE 14, 9
        PRINT "% COVERAGE =": CO
        LOCATE 16, 9
        PRINT "POPULATION ESTIMATE ="; PE
        PRINT 18, 9
        PRINT "POPULATION VARIANCE ="; VA#
        LOCATE 20, 9
        PRINT "STANDARD ERROR =" ; SE
        LOCATE 22, 9
        PRINT "95% CONFIDENCE LIMIT = +/-"; CL
        LOCATE 24, 9
        PRINT "% CONFIDENCE LIMIT = +/-"; PCL
        LOCATE 25, 9
        PRINT "M = MENU"
111:    GOSUB 50
```

```
        IF L2 = 77 THEN GOTO MENU
        IF NOT L2 = 77 THEN BEEP
        GOTO 111

METHOD 2:
                CLS
                A = 0
                B = 0
                S = 0
                AREA = " "
                SP2 = " "
                number = 0
                t20 = 8
                COLOR 7, 6
                LOCATE 4, 9
                PRINT "JOLLY'S METHOD 2 FOR UNEQUAL SIZED
                SAMPLE UNITS"
                LOCATE 5 , 9
                PRINT "-----------------------------------------------"
                LOCATE 7 , 19
                PRINT "Use BSPACE for correction ! ! ! !"
120:    LOCATE 9 , 9
        PRINT "Area of Survey Zone (km²) = ?"
        GOSUB 62
        A = X
        IF X = 0 THEN
        BEEP
        LOCATE 9, 45
        PRINT "     "
        t20 = t20 - 1
        GOTO 120
        END IF
121:    LOCATE 10 , 9
        PRINT "Number of Transects = ? "
        GOSUB 62
        number = X
        IF X < 2 OR X > 200 THEN
        BEEP
        LOCATE 10 , 37
        PRINT "     "
        LOCATE 10 , 45
        PRINT "> 1 AND < 200 transects"
```

```
        t20 = t20 - 1
        GOTO 121
        END IF
122:    LOCATE 11 , 9
        PRINT "Base-line length (km) = ?"
        GOSUB 62
        B = X
        IF X = 0 THEN
        BEEP
        LOCATE 11 , 45
        PRINT "     "
        t20 = t20 - 1
        GOTO 122
        END IF
123:    LOCATE 12, 9
        PRINT "Transect-width (m) = ?"
        GOSUB 62
        S = X / 1000
        IF X = 0 THEN
        BEEP
        LOCATE 12, 45
        PRINT "     "
        t20 = t20 - 1
        GOTO 123
        END IF

        LOCATE 13, 9
        PRINT "Conservation Area = ?"
        GOSUB 64
        AREA = SP1

        LOCATE 14, 9
        PRINT "Animal Species = ? "
        GOSUB 64
        SP2 = SP1

TUS6:   LOCATE 16, 9
        PRINT "DO YOU HAVE A CORRECTION FACTOR FOR
        THIS SPECIES,"
        LOCATE 17 , 9
        PRINT "OBTAINED THROUGH A DOUBLE-COUNT
        PROCEDURE. (Y)ES OR
```

```
         (N)O?"
         GOSUB 50
         IF L2 = 78 THEN CF = 1: GOTO TUS8
         IF L2 = 89 THEN GOTO TUS7
         IF NOT L2 = 78 OR L2 = 89 THEN BEEP: GOTO TUS6
TUS7:    t20 = 18
         LOCATE 19, 9
         PRINT "Correction factor = ? "
         GOSUB 62
         CF = X
         IF X < 1 THEN
         BEEP
         LOCATE 19, 35
         PRINT "     "
         t20 = t20 - 1
         GOTO TUS7
         END IF
TUS8:    LOCATE 24, 9
         PRINT "Press any key to continue"
         GOSUB 50

124:     CLS
         t4 = 0
         t5 = 0
         t6 = 0
         t7 = 0
         t8 = 0
         Tz = 0
         Tzy = 0
         Tz2# = 0
         Ty = 0
         Ty2# = 0
         K = 0
         K1 = 0
         K2 = 0
         K3 = 0
         K4 = 0
         K5 = 0
         K6 = 0
         K7 = 0
         K8 = 0
         K9 = 0
```

```
K10 = 0
K11 = 0
K12 = 0
K13 = 0
CO = 0
TUSSEN# = 0
R = 0
PE = 0
PE2 = 0
VA# = 0
SE = 0
CL = 0
PCL = 0
FOR t8 = 1 TO number
y(t8) = 0
TL (t8) = 0
Z (t8) = 0
Z2 (t8) = 0
Zy (t8) = 0
NEXT t8
t20 = 7
LOCATE 2, 9
PRINT "ENTER LENGTH OF EACH TRANSECT (KM);
(BSpace=correction)"
LOCATE 3 , 9
PRINT "----------------------------------------------------------------"
FOR t5 = 1 TO number
LOCATE 8 , 9
PRINT "Transect number" ; t5
GOSUB 62
TL (t5) = X
LOCATE 8 , 37
PRINT "     "
t20 = t20 - 1
NEXT t5
t20 = 18
LOCATE 13, 9
PRINT "ENTER NUMBER OF ANIMALS OBSERVED IN
EACH TRANSECT"
LOCATE 14 , 9
PRINT "----------------------------------------------------------------"
FOR t5 = 1 TO number
```

```
        LOCATE 19 , 9
        PRINT "Transect number"; t5
        GOSUB 62
        y (t5) = X
        LOCATE 8 , 37
        PRINT "     "
        t20 = t20 - 1
        NEXT t5
        t20 = 18
        LOCATE 13, 9
        PRINT "ENTER NUMBER OF ANIMALS OBSERVED IN
        EACH TRANSECT"
        LOCATE 14, 9
        PRINT "-----------------------------------------------------------------"
        FOR t5 = 1 TO number
        LOCATE 19 , 9
        PRINT "Transect number"; t5
        y (t5) = X
        LOCATE 19 , 37
        PRINT "-----------------------------------------------------------------"
        t20 = t20 – 1
        NEXT t5
        LOCATE 24 , 9
        PRINT "-----------------------------------------------------------------"
        LOCATE 25, 9
        PRINT "Press any key to continue"
        GOSUB 50
125:    t = 0
        IF number < 6 THEN t = 2 . 6
        IF number < 8 AND number > 5 THEN t = 2 . 4
        IF number < 10 AND number > 7 THEN t = 2 . 3
        IF number < 13 AND number > 9 THEN t = 2 . 2
        IF number < 27 AND number > 12 THEN t = 2 . 1
        IF number > 27 THEN t = 1. 9 6
129:    FOR t 4 = 1 TO number
        Z(t 4) = S * TL (t4)
        NEXT t4
        FOR t6 = 1 TO number
        Zy (t6) = y (t6) * Z (t6)
        Tzy = Tzy + Zy (t6)
        Ty = Ty + y (t6)
        Tz = Tz + Z (t6)
```

```
        NEXT t6
        FOR t7 = 1 TO number
        TUSSEN# = y (t7) © 2
        Ty2# = Ty2# + TUSSEN#
        Z2 (t7) = Z (t7) © 2
        Tz2# = Tz2# + Z2 (t7)
        TUSSEN# = 0
        NEXT t7
        R = Ty / Tz
        K = B / S
        REM Variance between animals counted
        K1 = 1 / (number - 1)
        K2 = Ty © 2
        K3 = Ty2# - (K2 / number)
        K4 = K1 * K3
        REM Variance between sample units
        K5 = Tz © 2
        K6 = Tz2# - (K5 / number)
        K7 = K1 * K6
        REM Covariance between animals counted and sample units
        K8 = (Ty * Tz) / number
        K9 = Tzy - K8
        K10 = K1 * K9
        REM Final calculations
        K11 = (K © 2 - K * number) / number
       PE2 = R * A
       PE = CF * PE2
       K12 = 2 * R * K10
       K13 = R © 2 * K7
       VA# = (CF © 2 ) * K11 * (K4 - K12 + K13)
       SE = SQR(VA#)
       CL = t * SE
       PE = CLNG (PE)
       SE = CLNG (SE)
       CL = CLNG (CL)
       CO = (Tz / A) * 100
       PCL = (CL / PE) * 100
140:    CLS
        LOCATE 4 , 35
        PRINT "RESULTS"
        LOCATE 5, 35
        PRINT "-------------"
```

```
        LOCATE 8 , 9
        PRINT "CONSERVATION AREA ="; AREA
        LOCATE 10, 9
        PRINT "ANIMAL SPECIES ="; SP2
        LOCATE 12, 9
        PRINT "CORRECTION FACTOR ="; CF
        LOCATE 14, 9
        PRINT "% COVERAGE ="; CO
        LOCATE 16, 9
        PRINT "POPULATION ESTIMATE ="; PE
        LOCATE 18 , 9
        PRINT "POPULATION VARIANCE ="; VA#
        LOCATE 20, 9
        PRINT "STANDARD ERROR ="; SE
        LOCATE 22, 9
        PRINT "95% CONFIDENCE LIMIT = +/-"; CL
        LOCATE 24, 9
        PRINT "% CONFIDENCE LIMIT = + / - "; PCL
        LOCATE 25 , 9
        PRINT "M = MENU"
141:      GOSUB 50
          IF L2 = 77 THEN GOTO MENU
          IF NOT L2 = 77 THEN BEEP
          GOTO 141

METHOD3:
              CLS
              t20 = 4
              AREA = " "
              SP2 = " "
              number = 0
              t5 = 0
              t6 = 0
              t7 = 0
              Tp = 0
              Tq = 0
              Tv = 0
              PE = 0
              VA# = 0
              SE = 0
              CL = 0
              PCL = 0
```

```
        LOCATE 2, 9
        PRINT "COMBINING THE RESULTS OF A
        STRATIFIED SAMPLE COUNT"
        LOCATE 3, 9
        PRINT "-------------------------------------------------------"
        LOCATE 5, 9
        PRINT "Conservation Area = ?"
        GOSUB 64
        AREA = SP1
        LOCATE 6, 9
        PRINT "Animal Species = ?"
        GOSUB 64
        SP2 = SP1

150:    t20 = 7
        LOCATE 8, 9
        PRINT "NUMBER OF STRATA = ?"
        GOSUB 62
        Number = X
        IF X > 20 THEN
        BEEP
        LOCATE 8, 37
        PRINT "    "
        LOCATE 8, 42
        PRINT "NOT MORE THAN 20"
        t20 = t20 - 1
        GOTO 150
        END IF
        FOR T2 = 1 TO number
        V (T2) = 0
        Q (T2) = 0
        P (T2) = 0
        NEXT T2
        t20 = 11
        LOCATE 10, 9
        PRINT "ENTER POPULATION ESTIMATES FOR EACH
        STRATUM"
        FOR t5 = 1 TO number
        LOCATE 12, 9
        PRINT "Stratum number" ; t5
        GOSUB 62
        P(t5) = X
```

```
        LOCATE 12, 37
        PRINT "     "
        t20 = t20 - 1
        NEXT t5
        t20 = 15
        LOCATE 14, 9
        PRINT "ENTER POPULATION VARIANCES FOR EACH STRATUM"
        FOR t6 = 1 TO number
        LOCATE 16, 9
        PRINT "Stratum number"; t6
        GOSUB 62
        Q(t6) = X
        LOCATE 16, 37
        PRINT "     "
        t20 = t20 - 1
        NEXT t6
        t20 = 19
        LOCATE 18 , 9
        PRINT "ENTER NUMBER OF TRANSECTS FOR EACH STRATUM"
        FOR t7 = 1 TO number
        LOCATE 20 , 9
        PRINT "Stratum number"; t7
        GOSUB 62
        V (t7) = X
        LOCATE 20, 37
        PRINT "     "
        t20 = t20 - 1
        NEXT t7
        LOCATE 24 , 9
        PRINT "Press any key to continue"
        GOSUB 50
160:    FOR t6 = 1 TO number
        Tp = Tp + P (t6)
        Tq = Tq + Q (t6)
        Tv = Tv + V (t6)
        NEXT t6
161:    t = 0
        IF Tv < 6 THEN t = 2.6
        IF Tv < 8 AND Tv > 5 THEN t = 2.4
        IF Tv < 10 AND Tv > 7 THEN t = 2.3
```

```
        IF Tv < 13 AND Tv > 9 THEN t = 2.2
        IF Tv < 27 AND Tv > 12 THEN t = 2.1
        IF Tv > 27 THEN t = 1.96
        PE = CLNG (Tp)
        VA# = CLNG (Tq)
        SE = SQR (VA#)
        SE = CLNG (SE)
        CL = t * SE
        CL = CLNG ( CL )
        PCL = ( CL / PE) * 100
162:    CLS
        LOCATE 4, 35
        PRINT "RESULTS"
        LOCATE 5, 35
        PRINT "-------"
        LOCATE 8, 9
        PRINT "CONSERVATION AREA ="; AREA
        LOCATE 10, 9
        PRINT "ANIMAL SPECIES ="; SP2
        LOCATE 12, 9
        PRINT "POPULATION ESTIMATE ="; PE
        LOCATE 14, 9
        PRINT "POPULATION VARIANCE ="; VA#
        LOCATE 16, 9
        PRINT "STANDARD ERROR =" ; SE
        LOCATE 18, 9
        PRINT " 95% CONFIDENCE LIMIT = +/-"; CL
        LOCATE 20 , 9
        PRINT " % CONFIDENCE LIMIT = +/-"; PCL
        LOCATE 25, 9
        PRINT "M = MENU"
163:    GOSUB 50
        IF L2 = 77 THEN GOTO MENU
        IF L2 < 77 AND L2 > 77 THEN BEEP
        IF L3 OR L4 THEN BEEP
        GOTO 163

METHOD 4:
                CLS
                t20 = 0
                P1 = 0
                P2 = 0
```

```
V1 = 0
V2 = 0
D = 0
P3 = 0
V3 = 0
LOCATE 2 , 15
PRINT "COMPARING TWO POPULATION ESTIMATES"
LOCATE 3, 9
PRINT "-------------------------------------------------------"
LOCATE 5, 9
PRINT "This section tests whether the difference between"
LOCATE 6, 9
PRINT "two population estimates is statistically significant."
LOCATE 7 , 9
PRINT "However, the number of sample units or transects should"
LOCATE 8, 9
PRINT "be larger than 30 for each estimate, while the samples"
LOCATE 9, 9
PRINT "should have been drawn independently from"
LOCATE 10, 9
PRINT "each other."
LOCATE 11, 9
PRINT "-------------------------------------------------------"
t20 = 12
LOCATE 13, 9
PRINT "POPULATION ESTIMATE 1 = ?"
GOSUB 62
P1 = X
t20 = 14
LOCATE 15, 9
PRINT "POPULATION VARIANCE 1 = ?"
GOSUB 62
V1 = X
t20 = 16
LOCATE 17, 9
PRINT "POPULATION ESTIMATE 2 = ?"
GOSUB 62
P2 = X
```

```
            t20 = 18
            LOCATE 19, 9
            PRINT "POPULATION VARIANCE 2 = ?"
            GOSUB 62
            V2 = X
            P3 = P1 – P2
            V3 = SQR (V1 + V2)
            D = P3 / V3
            IF D < 1.96 THEN GOTO 170
            IF D > 1.96 THEN GOTO 180
170:    LOCATE 21, 9
        PRINT "THE TWO ESTIMATES ARE NOT STATISTICALLY
        DIFFERENT"
        GOTO 190
180:    LOCATE 21, 9
        PRINT "THE TWO ESTIMATES ARE STATISTICALLY
        DIFFERENT AT P < 0.05"
190:    LOCATE 24, 9
        PRINT "M = MENU"
200:    GOSUB 50
        IF L2 = 77 THEN GOTO MENU
        IF L2 < 77 AND L2 > 77 THEN BEEP
        IF L3 OR L4 THEN BEEP
        GOTO 200

METHOD5:
            CLS
            t = 0
            t1 = 0
            T2 = 0
            t3 = 0
            t4 = 0
            t5 = 0
            VG = 0
            PG = 0
            MT = 0
            K = 0
            K1 = 0
            F = 0
            SV = 0
            W = 0
            number = 0
```

```
PE = 0
VA# = 0
SE = 0
CL = 0
PCL = 0
VG1 = 0
VG2T = 0
FOR t4 = 1 TO 10
VARI (t4) = 0
POP (t4) = 0
VG2 (t4) = 0
W (t4) = 0
M (t4) = 0
NEXT t4
LOCATE 2, 9
PRINT "MERGING TWO OR MORE INDEPENDENT POPULATION
ESTIMATES"
LOCATE 3, 9
PRINT "---------------------------------------------------------"
LOCATE 5, 9
PRINT "If the biases in the aerial survey technique are kept"
LOCATE 6, 9
PRINT "constant during all sample counts, two or more estimates"
LOCATE 7 , 9
PRINT "can be used to arrive at a single estimate with low"
LOCATE 8 , 9
PRINT "variance. However, one must first investigate whether"
LOCATE 9, 9
PRINT "the differences between the population estimates are"
LOCATE 10, 9
PRINT "due to sample error or to real differences such as"
LOCATE 11, 9
PRINT "migration. A variance ratio or F-test is applied to"
LOCATE 12 , 9
PRINT "the data, and only when this gives an insignificant"
LOCATE 13, 9
PRINT "result, the estimates can be merged!"
```

```
        LOCATE 15 , 9
        PRINT "This program will perform an F-test, and after it
        has"
        LOCATE 16, 9
        PRINT "been established that the observed differences
        between"
        LOCATE 17, 9
        PRINT "the population estimates are due to sample error
        alone,"
        LOCATE 18, 9
        PRINT "a maximum of 10 population estimates can be
        merged"
        LOCATE 24 , 9
        PRINT "Press any key to continue"
        GOSUB 50
300:    CLS
        number = 0
        LOCATE 2, 9
        PRINT "ENTER THE POPULATION ESTIMATES AND
        VARIANCES"
        LOCATE 3, 9
        PRINT "---------------------------------------------------------------"
        t20 = 5
305:    LOCATE 6 , 9
        PRINT "NUMBER OF ESTIMATES = ?"
        GOSUB 62
        number = X
        IF X > 10 THEN
        BEEP
        LOCATE 6, 45
        PRINT "NOT MORE THAN 10"
        LOCATE 6, 37
        PRINT "    "
        t20 = t20 - 1
        GOTO 305
        END IF
        t20 = 8
        FOR t1 = 1 TO number
        LOCATE 9, 9
        PRINT " POPULATION ESTIMATE"; t1
        GOSUB 62
        POP (t1) = X
```

```
        LOCATE 9 , 37
        PRINT "      "
        t20 = t20 - 1
        NEXT t1
        t20 = 11
        FOR T2 = 1 TO number
        LOCATE 12 , 9
        PRINT "POPULATION VARIANCE"; T2
        GOSUB 62
        VARI (T2) = X
        LOCATE 12 , 37
        PRINT "      "
        t20 = t20 - 1
        NEXT T2
        FOR t3 = 1 TO number
        VG = VG + VARI (t3)
        VG2 (t3) = VARI (t3) © 2
        VG2T = VG2T + VG2 (t3)
        NEXT t3
        VG1 = VG / number
        K = 1 / (number - 1)
        K1 = VG2T - (VG © 2 / number)
        SV = K * K1
        F = SV / VG1
        number = number - 1
        IF number = 1 THEN GOTO 319
        IF number = 2 THEN GOTO 320
        IF number = 3 THEN GOTO 321
        IF number = 4 THEN GOTO 322
        IF number = 5 THEN GOTO 323
        IF number = 6 THEN GOTO 324
        IF number = 7 THEN GOTO 325
        IF number = 8 THEN GOTO 326
        IF number = 9 THEN GOTO 327
319:    IF F < 161.4 THEN GOTO 350
        IF F > 161.4 THEN GOTO 340
320:    IF F < 19 THEN GOTO 350
        IF F > 19 THEN GOTO 340
321:    IF F < 9.28 THEN GOTO 350
        IF F > 9.28 THEN GOTO 340
322:    IF F < 6.39 THEN GOTO 350
        IF F > 6.39 THEN GOTO 340
```

```
323:  IF F < 5.05 THEN GOTO 350
      IF F > 5.05 THEN GOTO 340
324:  IF F < 4.28 THEN GOTO 350
      IF F > 4.28 THEN GOTO 340
325:  IF F < 3,79 THEN GOTO 350
      IF F > 3,79 THEN GOTO 340
326:  IF F < 3.44 THEN GOTO 350
      IF F > 3.44 THEN GOTO 340
327:  IF F < 3.18 THEN GOTO 350
      IF F > 3.18 THEN GOTO 340
340:  LOCATE 15 , 9
      PRINT "THE F-RATIO IS SIGNIFICANT, WHICH MEANS
      THAT THE"
      LOCATE 16 , 9
      PRINT "OBSERVED DIFFERENCES BETWEEN THE
      POPULATION ESTIMATES"
      LOCATE 17 , 9
      PRINT "ARE REAL DIFFERENCES , AS A RESULT OF
      MIGRATION OR"
      LOCATE 18, 9
      PRINT "OTHER UNKNOWN FACTORS. THE POPULATION
      ESTIMATES CANNOT"
      LOCATE 19 , 9
      PRINT "BE MERGED"
      LOCATE 24 , 9
      PRINT "M = MENU"
341:  GOSUB 50
      IF L2 = 77 THEN GOTO MENU
      IF L2 < 77 AND L2 > 77 THEN BEEP
      IF L3 OR L4 THEN BEEP
      GOTO 341
350:  LOCATE 15 , 9
      PRINT "THE F-RATIO IS INSIGNIFICANT"
      LOCATE 16 , 9
      PRINT "THE POPULATION ESTIMATES CAN BE MERGED"
      number = number + 1
      t = 0
      IF number < 6 THEN t = 2.6
      IF number < 8 AND number > 5 THEN t = 2.4
      IF number < 10 AND number > 7 THEN t = 2.3
      IF number = 10 THEN t = 2.2
      FOR t4 = 1 TO number
```

```
        W (t4) = 1 / VARI (t4)
        W = W + W (t4)
        NEXT t4
        FOR t5 = 1 TO number
        M (t5) = W (t5) * POP (t5)
        MT = MT + M (t5)
        NEXT t5
        PE = MT / W
        VA# = 1 / W
        SE = SQR (VA#)
        CL = t * SE
        PE = CLNG (PE)
        SE = CLNG (SE)
        CL = CLNG (CL)
        PCL = (CL / PE) * 100
        LOCATE 17 , 9
        PRINT "THE MERGED POPULATION ESTIMATE ="; PE
        LOCATE 18 , 9
        PRINT "THE VARIANCE ="; VA#
        LOCATE 19 , 9
        PRINT "THE STANDARD ERROR ="; SE
        LOCATE 20 , 9
        PRINT "THE 95% CONFIDENCE LIMIT = +/-"; CL
        LOCATE 21 , 9
        PRINT "THE % CONFIDENCE LIMIT = +/-"; PCL
        LOCATE 24 , 9
        PRINT "M = MENU"
360:    GOSUB 50
        IF L2 = 77 THEN GOTO MENU
        IF L2 < 77 AND L2 > 77 THEN BEEP
        IF L3 OR L4 THEN BEEP
        GOTO 360
HELP:
        CLS
        COLOR 7, 6
        LOCATE 2 , 33
        PRINT "HELP"
        LOCATE 4, 25
        PRINT "JOLLY'S METHODS 1 AND 2"
        LOCATE 5, 25
        PRINT "---------------------------------------"
        LOCATE 6, 9
```

```
PRINT "Jolly's Method 1 for equal sized sample units should"
LOCATE 7, 9
PRINT "be used when the sample units or transects are all of"
LOCATE 8, 9
PRINT "the same size. Jolly's Method 2 for unequal sized
sample"
LOCATE 9 , 9
PRINT "units should be used when the sample units or transects"
LOCATE 10 , 9
PRINT "have different sizes or lengths! When using the second"
LOCATE 11, 9
PRINT "method, however, you should make certain that the
length"
LOCATE 12 , 9
PRINT "of each transect can be measured accurately!"
LOCATE 14, 9
PRINT "Do not forget to sum the width of left and right"
LOCATE 15, 9
PRINT "strips in m, before you feed this information into the"
LOCATE 16, 9
PRINT "program! Note that the population estimate confidence"
LOCATE 17 , 9
PRINT "limits have been corrected for sample size"
LOCATE 18, 9
PRINT "by the program."
LOCATE 20, 9
PRINT "When the aerial survey included a double-count,"
LOCATE 21, 9
PRINT "visibility correction factors will be available"
LOCATE 22, 9
PRINT "for each species. The program gives you the option to"
LOCATE 23, 9
PRINT "enter the correction factor, after which the population"
LOCATE 24, 9
PRINT "statistics for that species will be revised accordingly."
LOCATE 25, 9
PRINT "Press any key to continue"
GOSUB 50
CLS
LOCATE 3, 25
PRINT "STRATIFIED SAMPLE COUNTS"
LOCATE 4, 25
```

```
PRINT "----------------------------------------"
LOCATE 6, 9
PRINT "If the survey area has been stratified and separate samples"
LOCATE 7, 9
PRINT "drawn in each of the strata, then each stratum is dealt with"
LOCATE 8, 9
PRINT "as a separate survey area. The program can be used to combine"
LOCATE 9, 9
PRINT "the strata to give a population estimate for the entire survey"
LOCATE 10, 9
PRINT "area"
LOCATE 12 , 25
PRINT "COMPARING ESTIMATES"
LOCATE 13, 25
PRINT "---------------------------------"
LOCATE 15, 9
PRINT "If you wish to investigate whether a population of animals changed"
LOCATE 16, 9
PRINT "in size, or whether two different methods give the same results,"
LOCATE 17, 9
PRINT "it is necessary to examine whether the difference is statistically"
LOCATE 18, 9
PRINT "different. The test provided in the program only applies if the"
LOCATE 19, 9
PRINT "number of transects counted is more than 30!"
LOCATE 24, 9
PRINT "Press any key to continue"
GOSUB 50
CLS
LOCATE 3, 25
PRINT "MERGING ESTIMATES"
LOCATE 4, 25
PRINT "----------------------------------------"
LOCATE 6, 9
```

```
        PRINT "When aerial sample counts are done at regular intervals,"
        LOCATE 7, 9
        PRINT "high variances of single estimates can be reduced by"
        LOCATE 8, 9
        PRINT "merging two or more estimates"
        LOCATE 24, 9
        PRINT "M = MENU"
550:    GOSUB 50
          IF L2 = 77 THEN GOTO MENU
          IF L2 < 77 AND L2 > 77 THEN BEEP
          IF L3 OR L4 THEN BEEP
          GOTO 550
FINISH:
          CLS
          COLOR 4, 7
          LOCATE 12, 25
          PRINT "HAVE A NICE DAY!"
          LOCATE 24, 9
          PRINT "Press any key to continue"
          GOSUB 50
          CLS : END
erro:
          IF ERR = 25 THEN em$ = "DEVICE FAULT!!": RESUME
          MENU
          IF ERR = 53 THEN em$ = "FILE NOT FOUND !!": RESUME
          MENU
          IF ERR = 53 OR ERR = 57 THEN em$ = "ERROR, WRONG
          DISC!": RESUME MENU
          IF ERR = 55 THEN em$ = "ERROR, TRY AGAIN!":
          RESUME erei
          IF ERR = 61 THEN em$ = "ERROR, DISC FULL!!":
          RESUME erei
          IF ERR = 68 THEN em$ = "DEVICE UNAVAILABLE!!":
          RESUME MENU
          IF ERR = 70 THEN em$ = "ERROR, DISC PROTECTED!!":
          RESUME erei
erei:   DEF SEG : RESET: GOTO MENU
OK
```

Index